PHYSICS RESEARCH AND TECHNOLOGY

INTERACTION OF THE ELECTROMAGNETIC FIELD WITH SUBSTANCE

PHYSICS RESEARCH AND TECHNOLOGY

Additional books in this series can be found on Nova's website under the Series tab.

Additional e-books in this series can be found on Nova's website under the e-book tab.

PHYSICS RESEARCH AND TECHNOLOGY

INTERACTION OF THE ELECTROMAGNETIC FIELD WITH SUBSTANCE

ANDREY NIKOLAEVICH VOLOBUEV

New York

For permission to use material from this book please contact us:
Telephone 631-231-7269; Fax 631-231-8175
Web Site: http://www.novapublishers.com

Library of Congress Cataloging-in-Publication Data

Volobuev, A. N. (Andrei Nikolaevich)
[Vzaimodeistvie vrashchaiushchikhsia elektromagnitnykh polei s zhivymi organizmami. English]
Interaction of the electromagnetic field with substance / Andrey Nikolaevich Volobuev (head of the chair - physics, Samara State Medical University, Samara, Russia).
pages cm
Includes index.
ISBN 978-1-62618-348-3 (soft cover)
1. Electromagnetism--Physiological effect. I. Title.
QP82.2.E43V6513 2013
612'.01442--dc23

2013008992

Published by Nova Science Publishers, Inc. † New York

CONTENTS

PREFACE

In this book, the problems of interaction of the electromagnetic field and substance are considered. This interaction is investigated on the basis of classical and quantum approaches. Standard methods, such as the classical electronic theory of interaction of the electromagnetic field with atoms, application of Maxwell equations to various problems and research of effective sections of interaction are discussed. Questions regarding the research of optical active substances, including the physical sense of the parameter of chirality and the relativistic theory of optical activity, are analyzed. Nonlinear aspects of the formation of electromagnetic impulse in dielectrics are submitted, including self-induced transparency. Decisions of some kinetic equations, including some for radiating gas, are given. On the basis of quantum electrodynamics, the quantum-relativistic approach to an explanation of Malus' law is developed, and some new features of deduction of the Klein-Nishina formula are investigated. This book will be useful to students and scientists working in areas connected with the interaction of electromagnetic radiation and substance.

INTRODUCTION

The existence of two forms of a matter of substance and field is the basic paradigm of physics. These forms of matter differ from each other, but the impenetrable border between them is not present. For example, in nuclear reactions there can be a transition of a matter in the energy of electromagnetic radiation. This transition is described by Einstein's well-known formula: $E = mc^2$.

For concretization of the problem, we shall concentrate our attention on the interaction of an electromagnetic field or radiation with substance. Each person meets this interaction everywhere; for example, including light in a room. Moreover, the basic data on the structure of a substance have been received by studying the influence of light on it.

Interaction of light and substance can be examined at different levels. These levels are determined in many respects by the depth of the description of the structure of substance or light.

It is necessary to distinguish between the classical and quantum approaches to the description of interaction of light and substance.

In the classical approach, light is always considered an electromagnetic wave. As to substance; the depth of its description can be completely varied. It is possible to examine the planetary model of atom, for example. It is convenient for the research of spectral effects of interaction of light with substance. It is possible to use so-called material equations. This approach is applied at the decision of Maxwell's equations, which will be discussed further below.

In the quantum approach, light is represented as a flux of separate energy portions of electromagnetic radiation - quantums or photons. Classical and quantum approaches frequently lead to the same or very similar results.

In chapters 1-3 concerning the classical consideration of a question of interaction of electromagnetic radiation with substance, for the purpose of decreasing the archaism of a statement, quantum phraseology is sometimes used. But actually, it is only phraseology since in these chapters there is no main position that distinguishes the quantum approach – the presence in the formal analysis of a constant of Planck.

It should be noted also that in considering the quantum and classical approaches, both from the point of view of depth of the analysis and from the point of view of its complexity, it is impossible to counter new with old as it is sometimes represented in poor quality textbooks. They reflect the real duality of our world, having arisen practically simultaneously and going back to the views, accordingly, of Newton and Huygens. It is simply that the quantum approach has undergone more development a little bit more recently than the classical approach.

I want to warn the reader that this book is not free from the author's own scientific interests, predilections and (frequently not trivial) approach to problems.

The author would like to thank the rector of Samara State Medical University academician RAMS, professor G.P. Kotelnicov, for his help in the preparation of the book for publishing.

This book will be useful to students and scientists working in areas connected with the interaction of electromagnetic radiation and substance.

We shall begin with consideration of the problem of interaction of light with substance on the basis of the classical approach.

Chapter 1

THE ELEMENTARY THEORY OF THE INTERACTION OF AN ELECTROMAGNETIC FIELD WITH A SUBSTANCE

1.1. THE QUALITATIVE ANALYSIS OF INTERACTION OF AN ELECTROMAGNETIC WAVE WITH ATOMS OF SUBSTANCE ON THE BASIS OF LINEAR CLASSICAL MODEL

Let's consider the interaction of light with atoms of substance. The preliminary analysis shall be carried out on the basis of the classical linear theory. Such analysis can give only general information on the effects of interaction, but the main principles of interaction will be cleared up. In this paragraph we shall reflect only on the qualitative analysis of the classical theory of interaction of light with atoms.

It is supposed that with a ray of light falling on a substance, energy of oscillations of electric field strength in a light wave is spent for the excitation of the forced oscillations of an electronic orbit of atoms around their nucleus. In the formula and are the current and maximal of electric field strength in an electromagnetic wave, - cyclic frequency of the forced oscillations, *t* - time.

As a result of the influence of an electromagnetic wave on an atom, there is an oscillating dipole - oscillator in which the centers of positive and negative charges are periodically interchanged in the position, figure 1.1.1.

The part of energy of a wave is transferred to the nucleus and also results in oscillations, i.e. actually passes in thermal energy and, hence, is absorbed. The other part is reradiated by an oscillating dipole as an electromagnetic wave which also passes deeper into the substance.

Oscillations of electric field strength in the reradiated wave look like $E_2 = E_{m2}\sin(\omega_f t - \Delta\varphi)$, where is a phase of delay of the forced oscillations of an electronic orbit concerning oscillations of compelling force $F_f = eE_1$, where *e* is a charge of electron.

As a result of absorption of light $E_{m2} < E_{m1}$

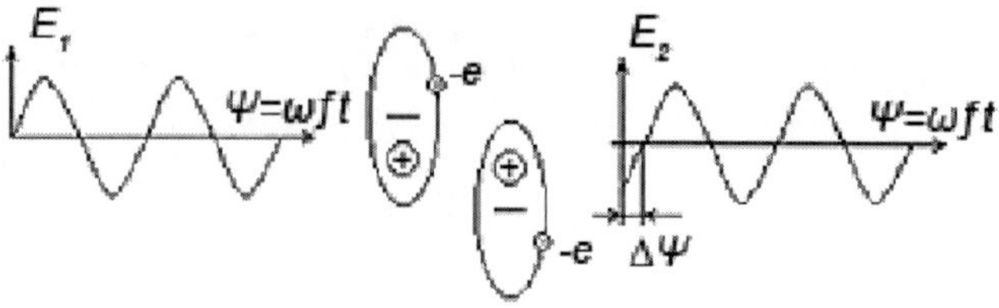

Figure 1.1.1.

From the considered classical picture of the interaction of electromagnetic oscillations and the atom, using the properties of the forced oscillations, it is possible to make the following three conclusions:

1. The electromagnetic wave radiated by an oscillating dipole has the frequency of the forced oscillations of the electronic orbit, equal to the frequency of the compelling force ω_f. I.e. at light propagation to the environment, the transition of light from a vacuum in a medium or from one medium to another, changes of frequency of light do not occur.
2. The phase of delay determines the reduction of the phase speed V of light in the environment in comparison with the speed of light in a vacuum c, i.e. $V < c$.
3. If a frequency of an electromagnetic wave falling on substance - frequency of the forced oscillations - is equal to the frequency of the oscillations of an electronic orbit of the atom, there is a resonance accompanied by a sharp increase of absorption of wave energy. Hence, the dependence of a parameter of light absorption k from the frequency of light appears as shown in figure 1.1.2. Values are frequencies of oscillations of electronic orbits of atoms of substance. This dependence determines a spectrum of absorption.

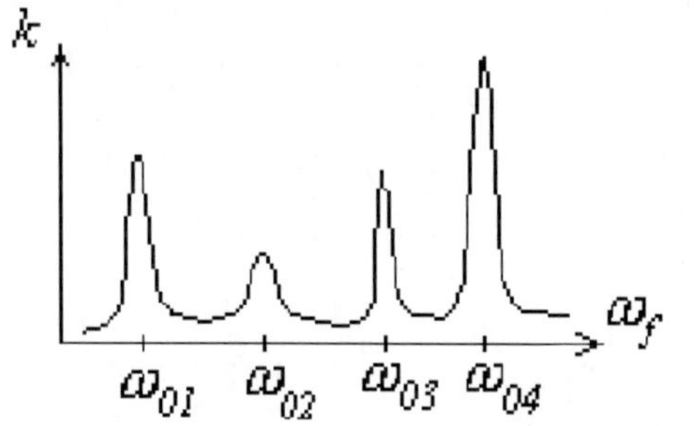

Figure 1.1.2.

In the field of "own frequencies" of oscillations of electronic orbits the phenomenon of an abnormal dispersion is observed.

The area in which the parameter of refraction grows with increasing frequency is referred to as an area of normal dispersion. The area in which the parameter of refraction decreases in frequency is referred to as an area of abnormal dispersion. The graphic dependence of a parameter of refraction of substance on the frequency of an electromagnetic wave falling on it is referred to as a dispersive curve. A typical dispersive curve is shown in figure 1.1.3.

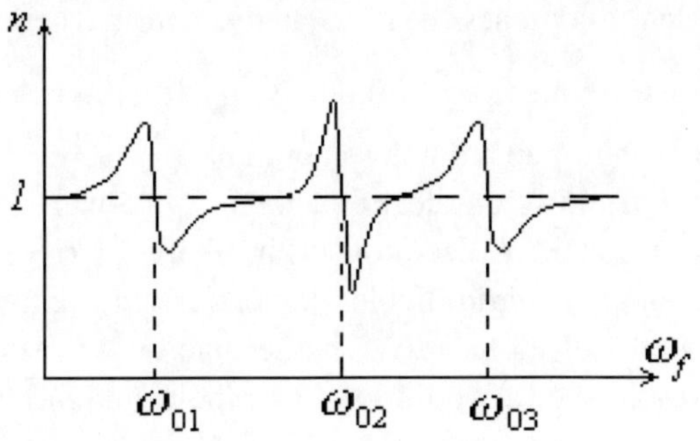

Figure 1.1.3.

In the field of an abnormal dispersion the phase of delay changes the sign, i.e. $E_2 = E_{m2}\sin(\omega_f t + \Delta\varphi)$, which results in an increase in the phase speed of light V in substance in comparison with the speed of light in vacuum c. If a frequency of a light wave exceeds little own frequency of oscillations of an electronic orbit of atom $\omega_f > \omega_{0i}$, there is an advancing on a phase by a light wave the oscillations of an electronic orbit which aspire to change their own frequency.

From the graph in figure 1.1.3, it can be seen that there are areas of frequencies in which the parameter of refraction is less than unit $n = \frac{c}{V} < 1$. This means that the phase speed of light in a substance is more than in a vacuum $V > c$, i.e. the vacuum appears optically denser than the substance. Such a situation occurs in the field of ultra-violet radiation for the substances (elements) which are taking place at the beginning of the periodic system of Mendeleev, and also for X-ray radiation and elements of an average part of the table of Mendeleev. It does not contradict the known position of Einstein's noted theory of relativity that the speed of light in a vacuum is the greatest possible. The group speed of a wave packet, i.e. the speed of carriage of energy by a wave packet, is always less than the speed of light in a vacuum.

We can note that from the quantum point of view, at the interaction of quantum of light with atom, at the approach of frequency of light to the area of a resonance, the aspiration of the atom to absorption of quantum grows.

It is expressed in the growth of effective section of absorption (see below) which starts to exceed the geometrical sizes of the atom considerably. The influence on quantum of atom area of absorption results in an increase in the phase speed of a light wave and, as consequence, the reduction of a parameter of refraction of substance with an increase in the frequency of light, i.e. to an abnormal dispersion.

1.2. Electronic Theory of Interaction of Light with Substance

We shall briefly consider the process of reradiated electromagnetic waves frequency ω by an electronic orbit with own frequency oscillation ω_0. It is known that according to Bohr the electron in a stationary orbit of an atom does not radiate.

If an oscillating electric field of an electromagnetic wave were to act on an atom, there would be oscillatory movements of an electronic orbit, figure 1.1.1.

The equation for oscillating electric dipole moment arisen oscillator, where e - a charge of electron, - the radius-vector of the positive charge (nucleus) directed from the center of a negative charge (an electronic shell), would look like the equation of the forced oscillations [1]:

$$\frac{d^2\boldsymbol{p}}{dt^2} + \gamma\frac{\partial \boldsymbol{p}}{\partial t} + \omega_0^2\boldsymbol{p} = \frac{e^2}{m}\boldsymbol{E}_0 e^{i\omega t}, \tag{1.2.1}$$

where m is the mass of electron, γ - characterizes attenuation of oscillations of an electronic orbit ($\frac{\gamma}{2}$ - factor of attenuation) and is the maximal size of electric field strength in an electromagnetic wave falling on the atom.

From the physical point of view, attenuation is caused by return power influence of the electromagnetic field radiated by electron on the electron, a so-called reaction of radiation or force of Lorentz's braking.

The decision of the equation (1.2.1) is well-known:

$$\boldsymbol{p} = \frac{\frac{e^2}{m}\boldsymbol{E}_0 e^{i\omega t}}{\omega_0^2 - \omega^2 + i\gamma\omega}. \tag{1.2.2}$$

It should be noted that with a phase delay of oscillations of the oscillator concerning oscillations of an electromagnetic wave falling on it which determines the delay of the speed of light in substance in comparison with that in a vacuum, it is possible to calculate using the formula:

$$\Delta\varphi = \text{arctg}\frac{\gamma\omega}{\omega_0^2 - \omega^2}. \tag{1.2.3}$$

On the basis of the decision of the equation (1.2.1) it is possible to estimate some parameters of interaction of an electromagnetic wave and substance.

As a result of influence on the substance of an external electric field, there is own electric field of substance $\boldsymbol{E}_c$, figure 1.2.1, directed against an external field $\boldsymbol{E}_0$. This field is also referred to as a field of coupled charges.

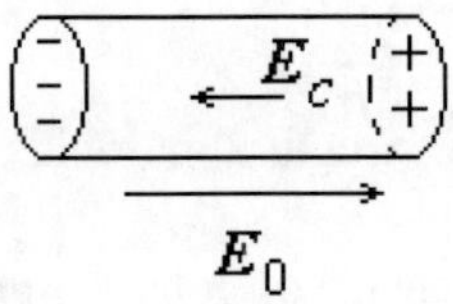

Figure 1.2.1.

Occurrence of own electric field is referred to as the polarization of substance.

Owing to polarization of substance, on its opposite sides there are superficial polarizing charges.

The resulting electric field strength in substance equals:

$$E = E_0 - E_c,$$

where the minus sign reflects various directions of vectors and $\boldsymbol{E}_c$.

We can divide both parts of equality into the size of a resulting field:

$$1 = \frac{E_0}{E} - \frac{E_c}{E}. \qquad (1.2.4)$$

The value is referred to as the relative dielectric penetrance of substance and is the main electric characteristic of substance. The relative dielectric penetrance of substance is shown in how much less the electric field strength is in a vacuum compared to the electric field strength created by the same charges in substance.

The size ε of crystals depends on a direction of crystallographic axes. For example, at ferroelectrics this size happens when it is very large. The value , named a relative dielectric susceptibility of substance, is also the electric characteristic of substance.

From the following formula (1.2.4), we can see that both these characteristics of substance are dimensionless and are coupled by the formula:

$$\varepsilon = 1 + \chi. \qquad (1.2.5)$$

The given relationship in the dimensionless form characterizes the balance of electric field strength in substance: external, own and resulting.

The result of the influence of the electric field strength on the substance can also be characterized by a vector of polarization (or polarization).

The vector of polarization is a total electric dipole moment of a unit of volume of a substance:

$$\boldsymbol{P} = \frac{\sum_i \boldsymbol{p}_i}{V}, \tag{1.2.6}$$

where is a dipole moment of a separate structural particle of substance; for example, atom, V - volume of substance.

Own electric field is coupled to a vector of polarization of substance a ratio:

$$\boldsymbol{P} = \varepsilon_0 \boldsymbol{E}_c = \varepsilon_0 \chi \boldsymbol{E}.$$

The dipole moments of structural particles of substance are proportional to the resulting electric field strength in substance:

$$\boldsymbol{p} = \varepsilon_0 \alpha \boldsymbol{E}, \tag{1.2.7}$$

where α is the polarizability of one particle and is an electric constant.

If the numerical concentration of atoms-oscillators in a homogeneous environment is the vector of polarization of this environment equals:

$$\boldsymbol{P} = n_0 \boldsymbol{p} = n_0 \varepsilon_0 \alpha \boldsymbol{E} = \varepsilon_0 \chi \boldsymbol{E}, \tag{1.2.8}$$

where is a relative dielectric susceptibility of substance.

Using (1.2.2), we shall find a vector of polarization of environment as:

$$\boldsymbol{P} = n_0 \boldsymbol{p} = \frac{\frac{e^2}{m} n_0 \boldsymbol{E}_0 e^{i\omega t}}{\omega_0^2 - \omega^2 + i\gamma\omega}. \tag{1.2.9}$$

Equating (1.2.8) and (1.2.9), with the account $\boldsymbol{E} = \boldsymbol{E}_0 e^{i\omega t}$, we shall find a relative dielectric susceptibility of substance:

$$\chi = \frac{e^2 n_0}{\varepsilon_0 m} \frac{1}{\left(\omega_0^2 - \omega^2 + i\gamma\omega\right)}. \tag{1.2.10}$$

Hence, relative dielectric penetrance of substance, according to (1.2.5), equals:

$$\varepsilon = 1 + \frac{e^2 n_0}{\varepsilon_0 m} \frac{1}{\left(\omega_0^2 - \omega^2 + i\gamma\omega\right)}. \tag{1.2.11}$$

Near to area of a resonance at it is possible to allow $\omega_0^2 - \omega^2 \approx 2\omega_0(\omega_0 - \omega)$. In this case the formula (1.2.11) will be transformed to a form more comprehensible for the quantum approach:

$$\varepsilon = 1 + \frac{e^2 n_0}{\varepsilon_0 m \omega_0} \frac{1}{\left(\omega_0 - \omega \ + i\frac{\gamma}{2}\right)} = 1 + \frac{4\pi r_0 n_0 c^2}{\omega_0} \frac{1}{\left(\omega_0 - \omega \ + i\frac{\gamma}{2}\right)}, \tag{1.2.12}$$

where is a classical radius of electron. From the physical point of view the classical radius of electron this distance from the center of electron on which potential energy of electron electric field is equal to its full energy $E = mc^2$. The formula is correct for rather rarefied substances (gas). In the case of more dense substances besides the located polarized particles, influencing against each other, it increases the total polarization of substance.

It is possible to show for such substances a relative dielectric susceptibility $\chi = \dfrac{\alpha n_0}{1 - \alpha n_0 / 3}$. Using (1.2.5) for dense substances, it is possible to receive the so-called law of Clausius - Mossotti [2]:

$$\frac{\varepsilon - 1}{\varepsilon + 2} = \frac{1}{3}\alpha n_0. \tag{1.2.13}$$

The external electric field acting on a substance is equal to the vector sum of resulting and own electric field $\boldsymbol{E}_0 = \boldsymbol{E} + \boldsymbol{E}_c$. By taking into account the coupling of an external and resulting field $\boldsymbol{E}_0 = \varepsilon \boldsymbol{E}$, and also a vector of polarization and own electric field in substance $\boldsymbol{E}_c = \dfrac{\boldsymbol{P}}{\varepsilon_0}$, we shall find $\varepsilon \boldsymbol{E} = \boldsymbol{E} + \dfrac{\boldsymbol{P}}{\varepsilon_0}$. Hence:

$$\boldsymbol{D} = \varepsilon\varepsilon_0 \boldsymbol{E} = \varepsilon_0 \boldsymbol{E} + \boldsymbol{P}, \tag{1.2.14}$$

where is a vector of an electric field induction (or a vector of electrical displacement).

1.2.1. A Dispersion of Relative Dielectric Penetrance

The dispersion of relative dielectric penetrance is a dependence of relative dielectric penetrance of substance on frequency. The kinds of polarization causing a dispersion of relative dielectric penetrance can be various: ionic polarization, orientational polarization, electronic polarization of atoms, etc. We shall reflect on electronic polarization and repolarization atoms under the action of an electromagnetic field which was considered in paragraph 1.1.

The formula (1.2.8) assumes that there are oscillations of an atom electronic orbit, i.e. its polarization and repolarization have time to follow the change of a field. It is correct only for low enough frequencies. In reality, in substance are observed relaxing processes which determine the delay of polarization of the substance concerning an electric field which is causing this polarization.

The phase of delay is determined, first of all, by the presence of attenuation of oscillations of an atom electronic orbit (1.2.3) and which depends on the frequency of a field. Therefore, for the analysis of the dispersive phenomena it is possible in the equation (1.2.1) of orbit oscillation to add two further aspects in the left part:

$$\frac{\partial \boldsymbol{p}}{\partial t}+\frac{\omega_0^2}{\gamma}\boldsymbol{p}=\frac{e^2}{\gamma m}\boldsymbol{E}_0 e^{i\omega t}, \tag{1.2.15}$$

The decision of this linear equation can be written thus; $\boldsymbol{p}=\frac{e^2\boldsymbol{E}_0}{m\omega_0^2}\frac{e^{i\omega t}}{(1+i\omega\tau)}$, where is the so-called time of a relaxation [2, 6]. Taking into account (1.2.8), we shall find a vector of polarization of substance as:

$$\boldsymbol{P}=n_0\boldsymbol{p}=\frac{e^2 n_0\boldsymbol{E}_0}{m\omega_0^2}\frac{e^{i\omega t}}{(1+i\omega\tau)}=\frac{e^2 n_0}{m\omega_0^2}\frac{\boldsymbol{E}}{(1+i\omega\tau)}. \tag{1.2.16}$$

Comparing (1.2.16) with (1.2.8), we find a relative dielectric susceptibility of substance $\chi=\frac{e^2 n_0}{\varepsilon_0 m\omega_0^2}\frac{1}{(1+i\omega\tau)}$. If we consider that at the big frequency of $\varepsilon\rightarrow\varepsilon_\infty$, instead of the formula (1.2.5) it is necessary to use the formula $\varepsilon=\varepsilon_\infty+\chi$.

In this case, relative dielectric penetrance (1.2.5) equals:

$$\varepsilon=\varepsilon_\infty+\chi=\varepsilon_\infty+\frac{e^2 n_0}{\varepsilon_0 m\omega_0^2}\frac{1}{(1+i\omega\tau)}. \tag{1.2.17}$$

At $\omega \to 0$, value $\varepsilon = \varepsilon_0 = \varepsilon_\infty + \frac{e^2 n_0}{\varepsilon_0 m \omega_0^2}$, hence $\varepsilon_0 - \varepsilon_\infty = \frac{e^2 n_0}{\varepsilon_0 m \omega_0^2}$. At $\omega \to \infty$, value $\varepsilon = \varepsilon_\infty$. Thus, (1.2.17) it is possible to copy as:

$$\varepsilon = \varepsilon_\infty + \frac{\varepsilon_0 - \varepsilon_\infty}{(1 + i\omega\tau)}. \tag{1.2.18}$$

Having separated in (1.2.18) the real part from an imaginary part, we shall find:

$$\varepsilon = \varepsilon_1 + i\varepsilon_2, \tag{1.2.19}$$

where

$$\varepsilon_1 = \varepsilon_\infty + \frac{\varepsilon_0 - \varepsilon_\infty}{1 + \omega^2 \tau^2} \tag{1.2.20}$$

characterizes dielectric properties of substance, and

$$\varepsilon_2 = \frac{(\varepsilon_0 - \varepsilon_\infty)\omega\tau}{1 + \omega^2 \tau^2} \tag{1.2.21}$$

characterizes conducting properties of substance, see below (2.2.16).

Formulas (1.2.18) - (1.2.21) are referred to as formulas of Debye.

If (1.2.19) is transformed to an angular representation of a complex number, we get: $\varepsilon = \sqrt{\varepsilon_1^2 + \varepsilon_2^2}\, \exp(i\delta)$.

The angle is referred to as the angle of dielectric losses and characterizes the dissipation of energy of an electromagnetic field in substance.

In figure 1.2.2 for example, dispersive curves for orientational polarization are shown (a curve 1 for ice; a curve 2 for water). The area of dispersion of the relative dielectric penetrance of ice lies in a more low-frequency area of a spectrum than that of water. This reveals that in a liquid substance the molecules are less coupled to surrounding molecules than in a solid body.

Therefore, water is more easily exposed orientationally to polarization and repolarization than ice. The area of dispersion of ice and water is basically in a radio-frequency range. For steam, where the freedom of movement of molecules is even greater, the characteristic dispersion frequency (frequency of an excess of a dispersive curve) is even higher and is equal - infra-red area of a spectrum.

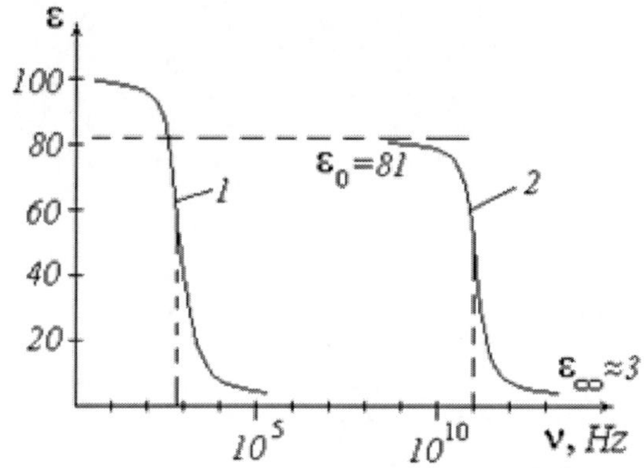

Figure 1.2.2.

It should be noted that proceeding from physical reasons, we have neglected the second derivative in the equation (1.2.15) and, investigating the kinetic equation of the first order, have received formulas of Debye for a dispersion of relative dielectric penetrance. Thus, we used the classical approach. But also, in using the quantum approach, in particular a principle detailed of balance, the kinetic equations of the first order are also applied. Thus, it is also possible to receive formulas of Debye (1.2.18) - (1.2.21) [4]. This testifies to the wide generality of these formulas.

Chapter 2

THE INTERACTION OF AN ELECTROMAGNETIC FIELD WITH SUBSTANCE: THE CLASSICAL APPROACH

2.1. MAXWELL EQUATIONS

Maxwell's equations represent the basic set (system) of equations for practical calculations in classical electrodynamics.

The physical sense of these equations consists, first of all, in the statement of continuity of an electromagnetic field.

For a conclusion of Maxwell's equations, using a whole standard approach, we shall briefly look at a conclusion of the continuity equation for the moving continuous medium.

According to the formula of Gauss – Ostrogradsky, a stream of mass Φ from some volume W (or the stream of a vector $\boldsymbol{j}$), limited to area $\boldsymbol{S}$, equals:

$$\Phi = \oint_{\boldsymbol{S}} \boldsymbol{j} d\boldsymbol{S} = \int_{W} \operatorname{div} \boldsymbol{j} dW, \tag{2.1.1}$$

where is a vector of density a stream of mass, - the module of a vector of density a stream of mass through the area S, $\boldsymbol{V}$ is a speed of movement of substance, ρ - its density, t - time.

By definition of a stream of mass we have:

$$\Phi = -\frac{\partial m}{\partial t} = -\frac{\partial}{\partial t} \int_{W} \rho dW = -\int_{W} \frac{\partial \rho}{\partial t} dW, \tag{2.1.2}$$

where the minus sign means a reduction of mass in volume W at the occurrence of a stream of mass Φ.

Equating (2.1.1) and (2.1.2), we find the continuity equation of the moving continuous medium:

$$\frac{\partial \rho}{\partial t} + \mathrm{div}\boldsymbol{j} = 0. \tag{2.1.3}$$

On the basis of consideration of a continuity of an electromagnetic field we shall put an integrated ratio [3]:

$$\boldsymbol{G} = \oint_{S} d\boldsymbol{S} \times \boldsymbol{E} = \int_{W} \mathrm{rot}\boldsymbol{E} dW, \tag{2.1.4}$$

where $\boldsymbol{E}$ is electric field strength. Vector $\boldsymbol{G}$ is determined by a ratio $\boldsymbol{G} = \oint_{S} d\boldsymbol{S} \times \boldsymbol{E}$.

Similarly (2.1.2) we shall enter a ratio:

$$\boldsymbol{G} = -\frac{\partial \boldsymbol{L}}{\partial t} = -\frac{\partial}{\partial t}\int_{W} \boldsymbol{B} dW = -\int_{W} \frac{\partial \boldsymbol{B}}{\partial t} dW, \tag{2.1.5}$$

where the vector is also similarly determined by mass in (2.1.2); $\boldsymbol{B}$ is induction of a magnetic field.

Equating (2.1.4) and (2.1.5), we find the first equation of a continuity of an electromagnetic field:

$$\frac{\partial \boldsymbol{B}}{\partial t} + \mathrm{rot}\boldsymbol{E} = 0. \tag{2.1.6}$$

The equation (2.1.6) represents one of Maxwell's equations for an electromagnetic field which expresses the Faraday law of electromagnetic induction.

Let's consider another of Maxwell's equations.

Let's write down a vector integrated ratio as:

$$\boldsymbol{N} = \oint_{S} d\boldsymbol{S} \times \boldsymbol{H} = \int_{W} \mathrm{rot}\boldsymbol{H} dW, \tag{2.1.7}$$

where $\boldsymbol{H}$ is magnetic field strength. Vector is connected to the sum of density of a current of charges $\boldsymbol{j}$ and a current of displacement $\boldsymbol{j}_{dis}$. It is determined by a ratio $\boldsymbol{N} = \oint_{S} d\boldsymbol{S} \times \boldsymbol{H}$.

Taking into account the definition of vector $\boldsymbol{N}$, we shall find:

$$\boldsymbol{N} = \int_{W} (\boldsymbol{j}_{CM} + \boldsymbol{j}) dW = \int_{W} \left(\frac{\partial \boldsymbol{D}}{\partial t} + \boldsymbol{j} \right) dW, \tag{2.1.8}$$

where is a vector; $\boldsymbol{D}$ is induction of an electric field.

Equating (2.1.7) and (2.1.8), we deduce the equation of Maxwell as:

$$\frac{\partial \boldsymbol{D}}{\partial t} + \boldsymbol{j} = \mathrm{rot}\boldsymbol{H}. \tag{2.1.9}$$

Thus, the continuity of an electromagnetic field is described by two of Maxwell's equations. In this relationship Maxwell's equations are completely opposite to the equations of quantum electrodynamics which basically consist in reflection quantum, i.e. the discontinuous nature of electromagnetic radiation. In its essence, it is also of the classical nature of Maxwell's equations.

The common physical sense of Maxwell's equations and, for example, the continuity equations of a stream of the continuous medium consists in the equivalence of positions: a change of a magnetic field generates a vortical electric field, and a change of mass in some volume generates a stream of mass from this volume. Both these positions reflect a continuity of a corresponding physical category. Together with the equations (2.1.6), (2.1.9), there are in addition two written equations.

The equation connecting an induction of electric field with volumetric density of free charges ρ:

$$\mathrm{div}\boldsymbol{D} = \rho. \tag{2.1.10}$$

The equation postulating absence of magnetic charges:

$$\mathrm{div}\boldsymbol{B} = 0. \tag{2.1.11}$$

The system of the equations (2.1.6), (2.1.9), (2.1.10) and (2.1.11) represents the system of Maxwell's equations. To these equations so-called material equations are added which, in their elementary form, look like this:

$$\boldsymbol{D} = \varepsilon\varepsilon_0\boldsymbol{E}, \tag{2.1.12}$$

$$\boldsymbol{B} = \mu\mu_0\boldsymbol{H}, \tag{2.1.13}$$

where is a magnetic constant and μ - relative magnetic permeability of substance.

With the addition of the material equations (2.1.12), (2.1.13) the system of Maxwell's equations becomes closed. Furthermore, we shall consider some applications of Maxwell's equations.

2.2. Application of Maxwell's Equations for Calculation of Penetration of Laser Radiation in Metal and Its Reflection from a Surface of a Metal Mirror

The processes of interaction between an electromagnetic wave and a metal surface represent the big interest at realization of various technological processes with the use of laser technique, and also at controlling the quality of the metal surface.

We shall consider a metal with a very pure surface (mirror) as a conducting, homogeneous and isotropic environment. The depth of penetration of an electromagnetic field in metal is very small. For example, on the depth equal to length of wave λ, the intensity of light decreases in one.

Maxwell's equations for electromagnetic fields in metal shall be written down as:

$$\mathrm{rot}\boldsymbol{H} = \boldsymbol{j} + \varepsilon_a \frac{\partial \boldsymbol{E}}{\partial t}, \tag{2.2.1}$$

$$\mathrm{rot}\boldsymbol{E} = -\mu_a \frac{\partial \boldsymbol{H}}{\partial t}, \tag{2.2.2}$$

$$\mathrm{div}\boldsymbol{E} = 0, \tag{2.2.3}$$

$$\mathrm{div}\boldsymbol{H} = 0, \tag{2.2.4}$$

where and are magnetic field strength and electric field strength in metal, - density of a current in the metal, arising under the action of an electric field, σ - specific electrical conductivity of metal, is absolute dielectric penetrance and - absolute magnetic permeability of metal. As a whole, we assume metal is electro-neutral; therefore, in (2.1.10) the volumetric density of free charges $\rho = 0$. We shall limit ourselves to consideration of flat, monochromatic (laser) waves.

First, we research the electromagnetic field penetrating into metal.

Let's find a rotor from the first equation:

$$\mathrm{rotrot}\boldsymbol{H} = -\Delta \boldsymbol{H} = \sigma \mathrm{rot}\boldsymbol{E} + \varepsilon_a \frac{\partial \mathrm{rot}\boldsymbol{E}}{\partial t}. \tag{2.2.5}$$

Substituting in (2.2.5) the equation (2.2.2), we shall find:

$$\Delta \boldsymbol{H} - \varepsilon_a \mu_a \frac{\partial^2 \boldsymbol{H}}{\partial t^2} = \sigma \mu_a \frac{\partial \boldsymbol{H}}{\partial t}. \tag{2.2.6}$$

Similarly:

$$\Delta \boldsymbol{E} - \varepsilon_a \mu_a \frac{\partial^2 \boldsymbol{E}}{\partial t^2} = \sigma \mu_a \frac{\partial \boldsymbol{E}}{\partial t}. \tag{2.2.7}$$

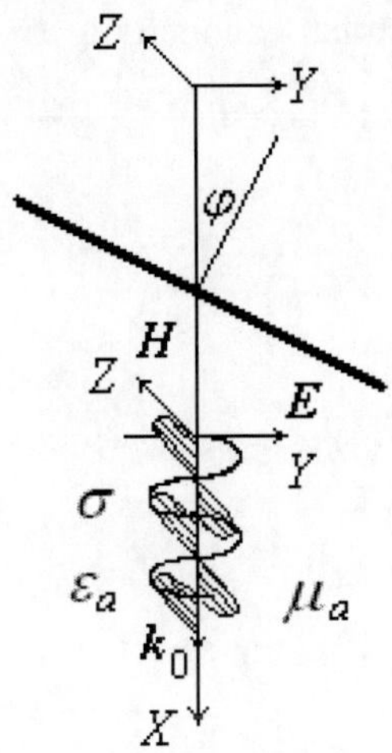

Figure 2.2.1.

Let's search for decisions of the wave equations (2.2.6) and (2.2.7) as flat attenuating monochromatic waves, figure 2.2.1, extending on axis X in metal:

$$\boldsymbol{H} = \boldsymbol{H}_0 \exp(-qX) \exp i(\omega t - pX), \tag{2.2.8}$$

$$\boldsymbol{E} = \boldsymbol{E}_0 \exp(-qX) \exp i(\omega t - pX), \tag{2.2.9}$$

where and are amplitudes of waves in metal, ω - cyclic frequency of waves, p - wave number, q - a parameter of absorption. Furthermore, we shall basically follow [4].

Substituting (2.2.8) and (2.2.9), accordingly, in (2.2.6) and (2.2.7) and equating the real and imaginary parts of the equations, we find:

$$p^2 - q^2 = \varepsilon_a \mu_a \omega^2, \tag{2.2.10}$$

$$2pq = \sigma \mu_a \omega. \tag{2.2.11}$$

Solving system of the equations (2.2.10) and (2.2.11), we find:

$$p = \omega \sqrt{\frac{\mu_a}{2}\left(\sqrt{\varepsilon_a^2 + \frac{\sigma^2}{\omega^2}} + \varepsilon_a\right)}, \tag{2.2.12}$$

$$q = \omega\sqrt{\frac{\mu_a}{2}\left(\sqrt{\varepsilon_a^2 + \frac{\sigma^2}{\omega^2}} - \varepsilon_a\right)}. \tag{2.2.13}$$

Let's consider a case where electrical conductivity is great in comparison with a current of displacement that is typical of metals, i.e. $\sigma >> \varepsilon_a\omega$.

Thus $\sqrt{\varepsilon_a^2 + \frac{\sigma^2}{\omega^2}} \approx \frac{\sigma}{\omega}$.

Hence $p \approx q \approx \omega\sqrt{\frac{\mu_a\sigma}{2\omega}} = \sqrt{\frac{\mu_a\sigma\omega}{2}}$. (2.2.14)

Let's enter complex dielectric penetrance of substance. According to (2.2.9), the derivative $\frac{\partial \boldsymbol{E}}{\partial t} = i\omega\boldsymbol{E}$.

The equation (2.2.1) will be transformed to a form:

$$\text{rot}\boldsymbol{H} = \sigma\boldsymbol{E} + \varepsilon_a i\omega\boldsymbol{E} = (\sigma + i\varepsilon_a\omega)\boldsymbol{E} = i\varepsilon^{/}\varepsilon_0\omega\boldsymbol{E} = \varepsilon^{/}\varepsilon_0\frac{\partial \boldsymbol{E}}{\partial t} =, \tag{2.2.15}$$

where is the complex relative dielectric penetrance of substance which takes into account its conducting properties. From (2.2.15) it follows that:

$$\varepsilon^{/} = \varepsilon - i\frac{\sigma}{\varepsilon_0\omega}. \tag{2.2.16}$$

Using a complex parameter of refraction of substance $n^{/} = \sqrt{\varepsilon^{/}\mu}$, we shall present it as:

$$n^{/} = n - ik, \tag{2.2.17}$$

where n is a parameter of refraction of substance, and k – the factor of absorption of electromagnetic radiation in substance.

Taking into account (2.2.16) and (2.2.17), we find:

$$\mu\varepsilon - i\frac{\mu\sigma}{\varepsilon_0\omega} = (n - ik)^2. \tag{2.2.18}$$

Equating the real and imaginary parts of the equation (2.2.18) and solving the received system as being relative to *n* and *k*, we shall find:

$$n = \frac{pc}{\omega} = \sqrt{\frac{\mu}{2\varepsilon_0}\left(\sqrt{\varepsilon_a^2 + \frac{\sigma^2}{\omega^2}} + \varepsilon_a\right)}, \qquad (2.2.19)$$

$$k = \frac{qc}{\omega} = \sqrt{\frac{\mu}{2\varepsilon_0}\left(\sqrt{\varepsilon_a^2 + \frac{\sigma^2}{\omega^2}} - \varepsilon_a\right)}, \qquad (2.2.20)$$

where is the speed of light in vacuum.

Big electrical conductivity of substances is typical of metal when $\sigma >> \varepsilon_a \omega$, it is possible to obtain:

$$n \approx k \approx \sqrt{\frac{\mu\sigma}{2\varepsilon_0\omega}}. \qquad (2.2.21)$$

Let's transform (2.2.17) to an angular representation of a complex parameter of refraction:

$$n' = \sqrt{\varepsilon'\mu} = \sqrt{n^2 + k^2}\exp(-i\delta), \qquad (2.2.22)$$

where the angle is frequently referred to as the angle of dielectric losses. The connection between the strength of a magnetic and electric field generally looks like [4]:

$$\boldsymbol{H} = \sqrt{\frac{\varepsilon'\varepsilon_0}{\mu\mu_0}}[\boldsymbol{k}_0 \times \boldsymbol{E}], \qquad (2.2.23)$$

where is a unit vector in the direction of spreading a wave.

Hence, substituting (2.2.22) in (2.2.23), we shall find:

$$\boldsymbol{H} = \sqrt{\frac{(n^2 + k^2)\varepsilon_0}{\mu^2\mu_0}}[\boldsymbol{k}_0 \times \boldsymbol{E}]\exp(-i\delta). \qquad (2.2.24)$$

Magnetic field strength in the absorbing (metal) environment phase lag from electric field strength on angle $\delta = \operatorname{arctg}\frac{k}{n}$. Let's consider the spreading of a flat wave, with polarization in planes *XY*, figure 2.2.1. The electric field strength (2.2.9) of a wave equals:

$$E_Y = E_0 \exp(-qX)\cos(\omega t - pX). \tag{2.2.25}$$

The magnetic field strength (2.2.24) of a wave equals:

$$H_Z = E_0\sqrt{\frac{(n^2+k^2)\varepsilon_0}{\mu^2\mu_0}}\exp(-qX)\cos(\omega t - pX - \delta). \tag{2.2.26}$$

The average time for the volumetric density of energy of an electromagnetic field can be found under the formula:

$$W = \frac{\varepsilon_a\langle E_Y^2\rangle}{2} + \frac{\mu_a\langle H_Z^2\rangle}{2}.$$

The mean square size of electric field strength (2.2.25) in a wave equals:

$$\langle E_Y^2\rangle = \frac{1}{T}\int_0^T E_Y^2 dt = \frac{1}{T}\int_0^T E_{Y0}^2 \exp(-2qX)\cos^2(\omega t - pX)dt = \frac{E_{Y0}^2}{2}\exp(-2qX)$$

The mean square size of magnetic field strength (2.2.26) in a wave equals:

$$\langle H_Z^2\rangle = \frac{1}{T}\int_0^T H_Z^2 dt = \frac{H_{Z0}^2}{2}\exp(-2qX).$$

Taking into account the average size of a square cosine in (2.2.25) and (2.2.26) is equal ½. Hence, the average time for volumetric density of energy of electromagnetic wave equals:

$$W = \frac{\varepsilon_a\langle E_Y^2\rangle}{2} + \frac{\mu_a\langle H_Z^2\rangle}{2} = \left(\frac{\varepsilon_a\langle E_{Y0}^2\rangle}{4} + \frac{\mu_a\langle H_{Z0}^2\rangle}{4}\right)\exp(-2qX)$$

Designating an average for the time volumetric density of energy of electromagnetic wave on coordinate as $W_0 = \frac{\varepsilon_a\langle E_{Y0}^2\rangle}{4} + \frac{\mu_a\langle H_{Z0}^2\rangle}{4}$, we shall find:

$$W = W_0 \exp(-2qX). \tag{2.2.27}$$

The ratio of the average time for the energy of magnetic and electric fields equals (coordinate indexes omitted):

$$\frac{\mu_a \langle H^2 \rangle}{\varepsilon_a \langle E^2 \rangle} = \frac{n^2 + k^2}{\varepsilon\mu} \approx \frac{\sigma}{\varepsilon_a \omega}. \tag{2.2.28}$$

At deduction (2.2.28), the formula (2.2.21) is taken into account, which is correct at $\sigma >> \varepsilon_a \omega$.

In metals, as previously noted, electrical conductivity is great; $\sigma >> \varepsilon_a \omega$; therefore, $\frac{\mu_a \langle H^2 \rangle}{\varepsilon_a \langle E^2 \rangle} >> 1$. Thus, the magnetic field of an electromagnetic wave will penetrate into metal much better than an electric field.

Let's consider now the process of reflection of electromagnetic radiation from a surface of metal.

The ratio of electric field strength in air to field strength in metal for a plane of polarization (a plane of oscillations of electric vector in a wave) a parallel plane of falling, according to the conditions of Fresnel, equals [5]:

$$\frac{E_{1\updownarrow}}{E_{2\updownarrow}} = -\frac{n^2 \cos\varphi - \sqrt{n^2 - (\sin\varphi)^2}}{n^2 \cos\varphi + \sqrt{n^2 - (\sin\varphi)^2}}. \tag{2.2.29}$$

where φ is the angle of falling of a light ray, figure 2.2.1.

The factor of reflection of a parallel component of a field is equal to a square of this ratio:

$$R_{\updownarrow} = \left(\frac{E_{1\updownarrow}}{E_{2\updownarrow}}\right)^2 = \left(\frac{n^2 \cos\varphi - \sqrt{n^2 - (\sin\varphi)^2}}{n^2 \cos\varphi + \sqrt{n^2 - (\sin\varphi)^2}}\right)^2. \tag{2.2.30}$$

We notice, according to (2.2.21), that the parameter of refraction for metal aspires to infinity if the electrical conductivity aspires to infinity. Therefore, we find a limit at aspiration $n \to \infty$. For this purpose having divided the numerator and the denominator by n^2, we obtain:

$$R_{\updownarrow} = \left(\frac{\cos\varphi - \sqrt{\frac{1}{n^2} - \frac{1}{n^4}(\sin\varphi)^2}}{\cos\varphi + \sqrt{\frac{1}{n^2} - \frac{1}{n^4}(\sin\varphi)^2}} \right)^2 = 1 \quad \text{at} \quad n \to \infty . \tag{2.2.31}$$

For a case of polarization, in a perpendicular plane of falling we have a factor of reflection:

$$R_{\leftrightarrow} = \left(\frac{E_{1\leftrightarrow}}{E_{2\leftrightarrow}} \right)^2 = \left(\frac{\cos\varphi - \sqrt{n^2 - (\sin\varphi)^2}}{\cos\varphi + \sqrt{n^2 - (\sin\varphi)^2}} \right)^2 . \tag{2.2.32}$$

Dividing the numerator and the denominator by n at aspiration $n \to \infty$, we shall find:

$$R_{\leftrightarrow} = \left(\frac{\frac{\cos\varphi}{n} - \sqrt{1 - \left(\frac{\sin\varphi}{n}\right)^2}}{\frac{\cos\varphi}{n} + \sqrt{1 - \left(\frac{\sin\varphi}{n}\right)^2}} \right)^2 = 1 \quad \text{at} \quad n \to \infty . \tag{2.2.33}$$

In spite of the fact that the electric field, as shown above, penetrates less well into metal than magnetic field, they are linked by formula (2.2.23). Therefore, if the electric field of electromagnetic wave does not penetrate into substance (metal), the magnetic field will not penetrate either. At and $\boldsymbol{H} = 0$.

Hence, in metal at the aspiration of electrical conductivity the factor of reflection to aspire to 1, i.e. the electromagnetic wave, is completely reflected. And this equally applies to electric and magnetic components waves.

The aspiration of electrical conductivity to infinity means that metal transforms in a superconducting state. It is possible at temperatures close to absolute zero on Calvin.

In this case, as shown in the analysis above, the phenomenon of superreflection is observed, i.e. falling under any angle to a surface of metal the light wave is completely reflected.

Let's consider the dependence of the factor of reflection on the length of a wave for a plane of polarization of a wave parallel to the plane of falling. For this purpose in the formula (2.2.30) we shall substitute (2.2.21). In the results, we observe:

$$R_{\updownarrow} = \left(\frac{\frac{\mu\sigma}{2\varepsilon_0\omega}\cos\varphi - \sqrt{\frac{\mu\sigma}{2\varepsilon_0\omega} - (\sin\varphi)^2}}{\frac{\mu\sigma}{2\varepsilon_0\omega}\cos\varphi + \sqrt{\frac{\mu\sigma}{2\varepsilon_0\omega} - (\sin\varphi)^2}} \right)^2 . \tag{2.2.34}$$

For a plane of polarization of a wave of a perpendicular plane of falling, substituting in (2.2.32) formula (2.2.21), we find:

$$R_{\leftrightarrow} = \left(\frac{\cos\varphi - \sqrt{\dfrac{\mu\sigma}{2\varepsilon_0\omega} - (\sin\varphi)^2}}{\cos\varphi + \sqrt{\dfrac{\mu\sigma}{2\varepsilon_0\omega} - (\sin\varphi)^2}} \right)^2 . \qquad (2.2.35)$$

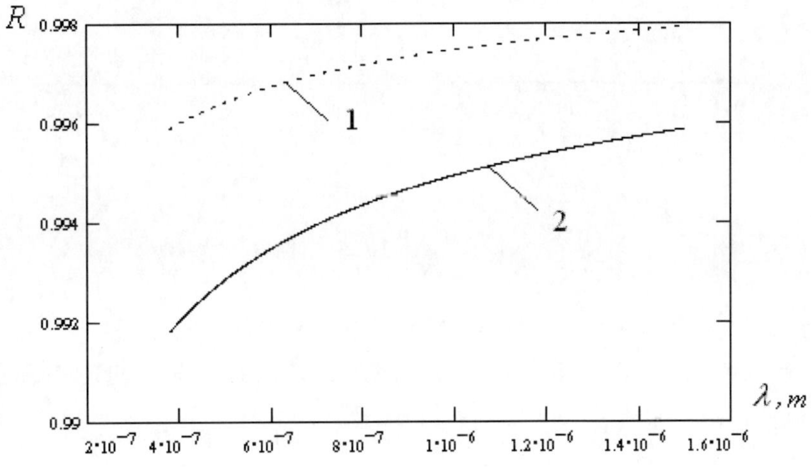

Figure 2.2.2.

In figure 2.2.2 the dependences of factors of reflection (a curve 1) and (a curve 2) from the wave-length which were calculated under formulas (2.2.35) and (2.2.34) in a range seen and near infra-red radiations are shown. As falling and reflected light spreads in air, the wave-length is calculated by the formula $\lambda = \dfrac{2\pi c}{\omega}$. With the increase in wave-length, factors of reflection grow, coming nearer to unit.

In figure 2.2.3 dependences of the factors of reflection (a curve 1) and (a curve 2) from specific electric conductivity of metal which increases at the reduction of temperature are shown.

The lowest sizes of electrical conductivity correspond to the electrical conductivity of steel at a temperature of $20^0 C$.

The wave-length was thus accepted as being equal to $\lambda = 632\ nm$, which corresponds to radiation of the red helium - neon laser. For both figures the falling angle of a ray was accepted as being equal to 45^0, relative magnetic permeability for steel $\mu \approx 5000$.

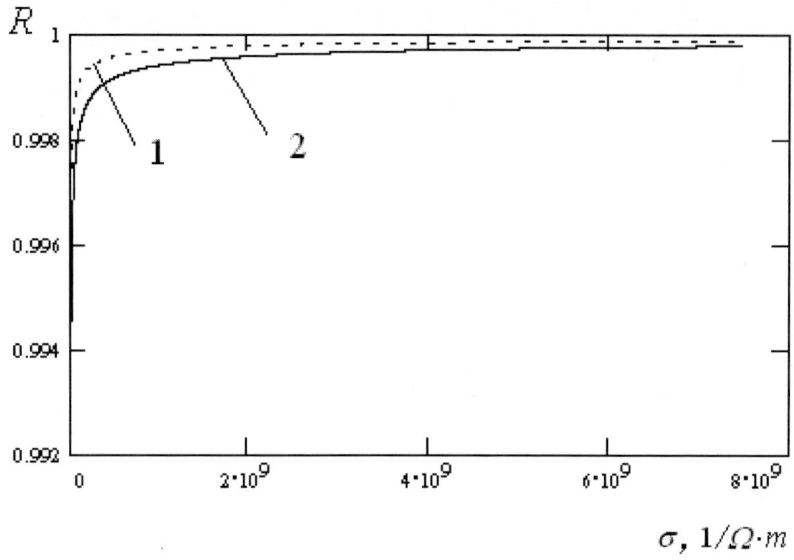

Figure 2.2.3.

2.3. Application of Maxwell's Equations for Calculation of Procedures of High-Frequency Physiotherapy in Medicine

One of the major directions for the treatment of patients, connected first of all with the heating up of the tissues of organisms, is action on the patient's high-frequency electromagnetic fields.

The physiatrist's understanding of the principles of the thermal influence of electromagnetic fields on human organisms, the skills to analyze the features of a thermal emission in medical procedures in various tissues of organism and the correct choice of this or that thermal high-frequency procedure adequate to the therapeutic purposes, are the major requirements for the successful treatment of the patient.

The purpose of the given paragraph is the comparative analysis of the features of high-frequency physiotherapeutic procedures: microwave therapy (MW-therapy), inductothermal and ultrahigh-frequency therapy (UHF - therapy) from the point of view of thermal effect.

2.3.1. Features of the Influence of the MW - Radiations on a Tissue of Organism

Wide application of microwave therapy in medicine is basically connected to the thermal influence of fields of the MW - range, i.e. a range of electromagnetic waves; the lengths of which lie in the interval $\lambda = 10\,mm \div 1\,m$.

It is important to correctly estimate the features of influence of electromagnetic radiation of the MW - range on tissues of organisms and in particular the thermal effect at MW - therapies.

Let's consider the absorption of the MW - radiations of small intensity in tissues of organisms. The absorption of electromagnetic radiation in tissues is determined by Bouguer's law:

$$\frac{\Delta I}{\Delta X} = -\alpha I, \tag{2.3.1}$$

where is the thickness of an absorbing layer, coordinate X is directed in the depth of a body of a man, α - a parameter of absorption of electromagnetic waves, I - intensity of electromagnetic radiation to tissues, - intensity of electromagnetic radiation falling on a tissue, figure 2.3.1.

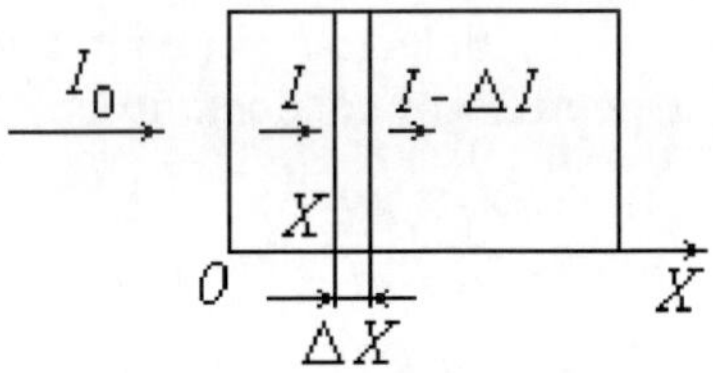

Figure 2.3.1.

Bouguer's law can be received in various ways. In particular, it follows from the decisions of the system of Maxwell's equations (2.2.1) - (2.2.4) which we shall use at the analysis of the given problem.

We shall find the average time for the volumetric density of energy of electromagnetic field, using (2.2.27) $W = W_0 \exp(-2qX)$. The intensity of a wave is coupled with W under the formula $I = WV$, where V is the speed of electromagnetic wave in the examined environment. Hence, the intensity of a wave equals:

$$I = I_0 \exp(-\alpha X), \tag{2.3.2}$$

where is the intensity of a wave on coordinate $X = 0$, a parameter of absorption (q is the designation of a parameter of the absorption, used earlier in paragraph 2.2). Bouguer's law (2.3.2) is shown in an integrated form. The solution to Bouguer's law is in the differential form (2.3.1) as $\frac{dI}{dX} = -\alpha I$.

In the given paragraph we use a designation q - the quantity of heat allocated in the unit of time in the unit of volume of substance on coordinate X. According to Bouguer's law (2.3.1), this quantity of heat equals:

$$q = -\frac{\Delta I}{\Delta X} = \alpha I . \tag{2.3.3}$$

The minus sign specifies that the reduction of the intensity of radiation in the environment is connected to an increase in its heat production q. The parameter of absorption of medium for electromagnetic radiation, using the formula (2.2.12), is complex function absolute dielectric penetrance and absolute magnetic permeability of medium, its specific electrical conductivity σ, and frequencies of electromagnetic radiation ω:

$$\alpha = \omega \sqrt{2\mu_a \left(\sqrt{\varepsilon_a^2 + \frac{\sigma^2}{\omega^2}} - \varepsilon_a \right)} . \tag{2.3.4}$$

Tissues of an organism represent compositing material consisting of a component, having electro-conducting (blood, muscles) and dielectric (skin, fat, bones) properties.

Let's consider a case when the current of conductivity in tissues is greater in comparison with a current of displacement, i.e. $\sigma >> \varepsilon_a \omega$.

Thus $\sqrt{\varepsilon_a^2 + \frac{\sigma^2}{\omega^2}} \approx \frac{\sigma}{\omega}$. Hence:

$$\alpha \approx \sqrt{2\mu_a \sigma \omega} . \tag{2.3.5}$$

According to (2.3.2) there is an intensity I on the depth and electric field strength E on depth decreases in one.

In tissues of organisms there are no ferromagnetic materials; therefore, it is possible to assume that the relative magnetic permeability of the tissue is $\mu \approx 1$. Hence, absolute magnetic permeability of the tissue is equal to a magnetic constant and the formula (2.3.5) becomes:

$$\alpha \approx \sqrt{2\mu_0 \sigma \omega} . \tag{2.3.6}$$

The quantity of heat allocated in unit of time in unit of volume of electro-conducting tissue on coordinate X equals:

$$q = \alpha I = k_1 \sqrt{2\mu_0 \sigma \omega} I , \tag{2.3.7}$$

where is the factor of proportionality which takes into account losses of electromagnetic radiation to surrounding space. Usually there is $k_1 \approx 0{,}7$.

Let's find the thermal effect in the organism's tissues with dielectric properties. The thermal effect in such tissues depends on a dielectric relationship and electro-conducting properties. For a definite result we shall assume that dielectric properties of a tissue in β are more expressed than electro-conducting properties:

$$\varepsilon_a \approx \beta \frac{\sigma}{\omega}. \tag{2.3.8}$$

The value $\frac{1}{\beta} = \mathrm{tg}\delta$, where δ is an angle of dielectric losses [6] describing the relationship between imaginary and real components of relative dielectric penetrance of the tissue, see (2.2.16).

From the formula (2.3.4) we shall find a parameter of absorption:

$$\alpha = \omega\sqrt{2\mu_a\varepsilon_a\left(\sqrt{1+\frac{1}{\beta^2}}-1\right)} = \omega\sqrt{2\mu_a\varepsilon_a\left(\frac{1}{\cos\delta}-1\right)}. \tag{2.3.9}$$

The quantity of heat allocated in a unit of time and in a unit of volume of a dielectric tissue on coordinate X equals:

$$q = \alpha I = k_2\omega\sqrt{\varepsilon}I. \tag{2.3.10}$$

With the account $\mu \approx 1$, we find the factor $k_2 \approx k_1\sqrt{2\mu_0\varepsilon_0\left(\frac{1}{\cos\delta}-1\right)}$. The factor takes into account the constants which are included in the formula (2.3.10) for parameter of absorption, such as a magnetic constant μ_0, an electric constant $\varepsilon_0 = 8.85\cdot10^{-12}\frac{F}{m}$, and also an angle of dielectric losses δ of the tissue and loss of energy of electromagnetic radiation to surrounding space k_1. Taking into account - the speed of light in vacuum, it is possible to write down $k_2 \approx \frac{k_1}{c}\sqrt{2\left(\frac{1}{\cos\delta}-1\right)} = \frac{2k_1\sin\delta/2}{c\sqrt{\cos\delta}}$. The inverse dependence of factor on speed of light in vacuum specifies small absorbing ability of dielectric tissues of organism.

Thus, the intensity of the MW - radiations in the tissues of an organism can be represented by the exponential law (2.3.2).

The parameter of absorption α grows with the increase in the frequency of radiation ω (2.3.9); therefore, long waves will penetrate more deeply into the tissues of the organism.

In figure 2.3.2 the dependences of the relative intensity of radiation designed on the formula (2.3.10) on depth of penetration in electro-conducting tissue X at CMW - therapy (a curve 1) and at DMW - therapy (a curve 2) are shown. CMW - therapy is centimeter microwave therapy, DMW - therapy is decimeter microwave therapy.

For calculation purposes, it is accepted that: specific electrical conductivity of tissue in area CMW - therapies (for a skeletal muscle) and in area DMW - therapies [7].

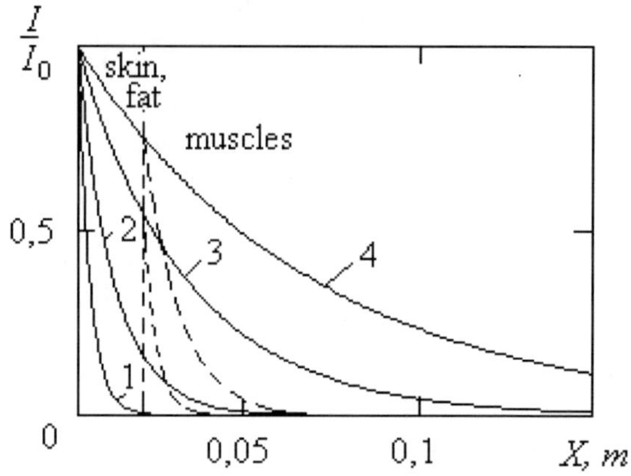

Figure 2.3.2.

There is cyclic frequency of electromagnetic field influence for CMW - therapy and for DMW - therapy. The parameter of absorption designed under the formula (2.3.6) for CMW - therapy and for DMW - therapy $\alpha = 93.3\, m^{-1}$.

From figure 2.3.2 it is clear that the depth of penetration of electromagnetic radiation in electro-conducting tissues of organisms is less than 2 *cm* at CMW - therapy and ~ 5 *cm* at DMW - therapy. The ratio of the depths of penetration is inversely proportional to the ratio of parameters at CMW and DMW - therapies:

$$\sqrt{\frac{(\sigma\omega)_{\text{CMW}}}{(\sigma\omega)_{\text{DMW}}}} = 2.93 \cdot$$

In figure 2.3.2 the diagrams of dependence of radiation intensity from depth of penetration into the fatty tissue, having expressed dielectric properties, are also shown: at CMW - therapy (a curve 3) and at DMW - therapy (a curve 4). For calculation purposes, it is accepted that: an angle of dielectric losses for a fatty tissue at CMW – therapy and at DMW – therapy $\delta = 37.7^0$. There is relative dielectric penetrance at CMW – therapy $\varepsilon = 6.5$, at DMW – therapy [7]. There is a parameter of absorption for CMB - therapy $\alpha = 31\, m^{-1}$, for DMW - therapy $\alpha = 14.5\, m^{-1}$.

The parameter of absorption for a dielectric tissue is calculated under the formula:

$$\alpha = \frac{2\sin\delta/2}{c\sqrt{\cos\delta}}\omega\sqrt{\varepsilon} \cdot \qquad (2.3.11)$$

Actually, the depth of penetration of the tissues of an organism varies in the beginning under the law corresponding to curves 3 and 4, since superficial layers (skin, fat) have appreciably dielectric properties where the absorption of electromagnetic radiation is less. Then, in a muscular layer there will be sharper absorption of energy of the MW - radiations approximately under the law corresponding to curves 1 and 2. The dotted line shows the law of absorption in a skeletal muscle for CMW - and DMW - therapies, conditionally transferred in anatomic area of a muscular layer.

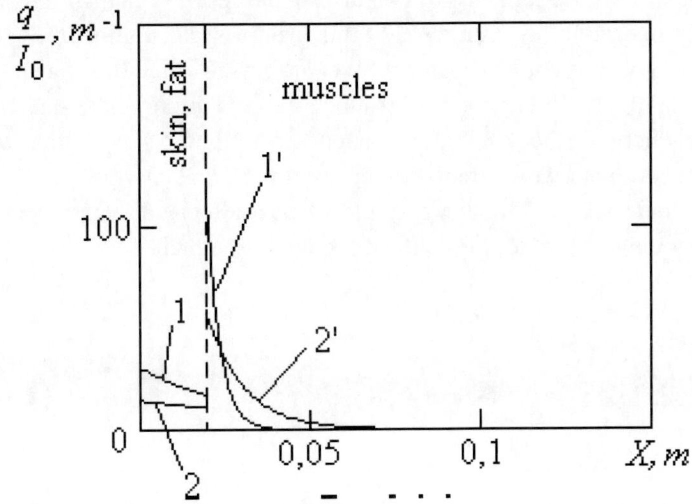

Figure 2.3.3.

Fig. 2.3.3 shows the result of the calculation of dependence of the relative thermal effect designed under the formula, following (2.3.1), from the depth of a tissue of an organism:

$$\frac{q}{I_0} = -\frac{1}{I_0}\frac{dI}{dX} \quad (2.3.12)$$

Actually, a sharp variation of heat production on the border between fat and muscle is not present, since anatomically, the transition occurs smoothly. The calculation shows that most of the heat is allocated to a muscular layer, directly adjoining a fatty layer. The skin and fat are heated up much less than muscles. As a whole, the heating of tissues at CMW - therapy (curves 1-1') is greater than at the DMW - therapy (curves 2-2'), especially in the field of transition to a muscular layer. But at the DMW – therapy, deep laying tissues are heated up more.

2.3.2. Features of Influence UHF - Radiations on a Tissue of Organism

Electromagnetic fields with ultrahigh ($30 \div 300 MHz$) frequencies (UHF) are used in medicine for heating tissues of organisms. The area of a body is located between two flat isolated electrodes which do not touch the body, i.e. actually between the capacitor plates. In the tissues which are placed in the oscillating electromagnetic field, due to an electric component of a field, there can be conductivity and displacement currents, depending on their structure.

2.3.2.1. UHF - Heating up of Electro-Conducting Tissues

Let's consider the allocation of heat at UHF - therapy in the electro-conducting tissue placed in the oscillating electric field.

The oscillating electric field in electro-conducting tissues of organisms creates alternating electric current *i*, proportional to strength *E* of this electric field, figure 2.3.4. The energy of an ultrahigh-frequency current *i*, basically, according to the Biot-Savart-Laplace law transforms to magnetic field energy with induction *B*. Furthermore, due to the law of electromagnetic induction, the oscillating magnetic field *B* causes the occurrence of alternating vortical currents *I* which transform energy to thermal energy.

According of Joule - Lenz law, the quantity of heat allocated in the electro-conducting tissue in a unit of volume in a unit of time (heat production) equals:

$$q = k_1 \frac{E_{eff}^2}{\rho}. \tag{2.3.13}$$

where ρ is the specific resistance of the medium.

In the formula for heat production q oscillating value of electric field strength $E = E_{\max} \sin \omega t$, is replaced with the effective value equivalent on thermal action. The factor reflects losses of energy of electromagnetic field on radiation in the surrounding space.

Let's consider the thermal effect at UHF - heating of electro-conducting tissues in more detail. Electro-conducting tissues of organisms show the tendency to shield it from the electromagnetic field acting on it. Alternating vortical currents *I* in internal areas of heated up body are directed against an external current *i* and compensate with an external current, as seen in figure 2.3.4. In external areas, the heated up body vortical currents *I* and the external current *i* are directed equally; that increases the common current. Therefore, the current and an electromagnetic field at UHF - therapy decreases at the internal areas of the body which complicates their heating process in this physiotherapeutic procedure.

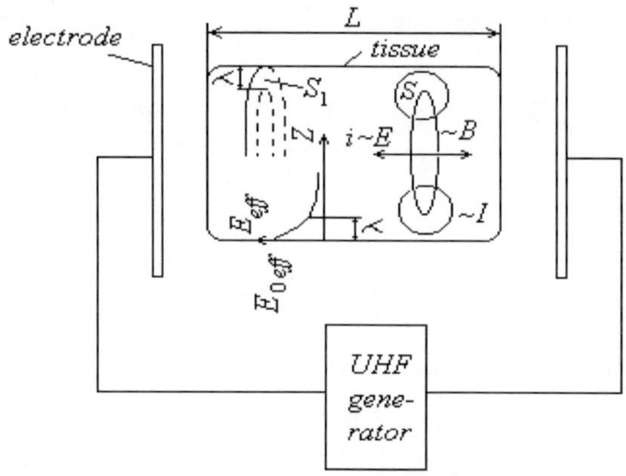

Figure 2.3.4.

To find the depth of penetration of an electric field in the electro-conducting tissue of an organism, we shall write down Maxwell's equations for such a tissue as:

$$\mathrm{rot}\boldsymbol{H} = \boldsymbol{j}, \tag{2.3.14}$$

$$\mathrm{rot}\boldsymbol{E} = -\mu_a \frac{\partial \boldsymbol{H}}{\partial t}, \tag{2.3.15}$$

$$\mathrm{div}\boldsymbol{E} = 0, \tag{2.3.16}$$

$$\mathrm{div}\boldsymbol{H} = 0, \tag{2.3.17}$$

where is the density of current *i*, and - magnetic and electric fields' strength. In equation (2.3.14) we take into account that the occurrence of a magnetic field occurs due to the alternating density of a current $\boldsymbol{j}$.

Finding a rotor from the equation (2.3.15) with the account (2.3.16), we obtain:

$$\mathrm{rotrot}\boldsymbol{E} = \mathrm{graddiv}\,\boldsymbol{E} - \Delta\boldsymbol{E} = -\Delta\boldsymbol{E} = -\mu_a \frac{\partial \mathrm{rot}\boldsymbol{H}}{\partial t}. \tag{2.3.18}$$

Using (2.3.14), gives us:

$$\frac{\partial^2 E}{\partial Z^2} = \mu_a \frac{\partial j}{\partial t} = \frac{\mu_a}{\rho}\frac{\partial E}{\partial t}. \tag{2.3.19}$$

where *Z* is the coordinate directed deep into a tissue, as in fig. 2.3.4. The basic gradient of the module of an electric field is directed on axis Z.

As opposed to the wave equations (2.2.6) and (2.2.7), which are the equations of the hyperbolic type, the equation (2.3.19) is of the parabolic type.

A physically comprehensible solution of the equation (2.3.19) looks like this:

$$E = E_0 \mathrm{e}^{-\frac{Z}{\lambda}} \mathrm{e}^{i\left(\omega t - \frac{Z}{\lambda}\right)}. \tag{2.3.20}$$

where is the oscillating value of the electric field strength on a surface of tissue at $Z = 0$. The last factor in (2.3.20) reflects the oscillatory character of a field. Value λ characterizes the depth of penetration of electromagnetic field in a tissue; known as a skin-layer. Substituting (2.3.20) in (2.3.19), we find that the size of skin-layer λ depends on the frequency of the electromagnetic field and the specific resistance of medium $\lambda = \sqrt{\frac{2\rho}{\mu_a \omega}}$.

The less specific the resistance of the medium and the greater the frequency of the electromagnetic field, the less the depth of penetration of an electric field into the medium and the less the depth of heating of electro-conducting tissue.

It should be noted that the thickness of a skin-layer in accuracy corresponds to the depth of penetration of the electric field in electro-conducting tissue on it's MW - radiations, see (2.3.5).

According to (2.3.20), the dependence of the effective size of electric field strength on the depth of its penetration in electro-conducting tissue, under conditions of its electric homogeneity, has an exponential character and looks like this:

$$E_{eff} = E_{0eff} \mathrm{e}^{-\frac{Z}{\lambda}}, \tag{2.3.21}$$

where is the effective value of the electric field strength on a surface of a tissue at $Z = 0$.

Due to the polarization of electro-conducting tissues at redistribution in the electric field of free ions, there is a superficial layer of absorption of the electromagnetic field's energy thickness λ and in the face parts of the heated up volume of the tissue, figure 2.3.4.

The decision of a task for face zones is similar to the aforementioned findings of the thickness of the skin-layer [8]. Shielding of the electromagnetic field is carried out due to the suppression of an electric component of the field. Penetration of the electromagnetic field into depth electro-conducting tissue occurs, on the contrary, due to a magnetic component of a field. Therefore, in an electro-conducting tissue, the energy of the electromagnetic field is concentrated basically in a magnetic component of the field. Let's consider in more detail a thermal emission at the UHF - therapy in electro-conducting tissues. The vortical current is equal to $I = \frac{\varepsilon}{R}$, where is the electromotive force of electromagnetic induction which is proportional to the speed of variance of the magnetic flux Φ, and R - resistance of the tissue. Taking into account that at the perpendicularity of a contour of the vortical current limited to area S, and a vector of an induction of a magnetic field B, a magnetic flux is $\Phi = BS$, we find:

$$I = -S\frac{dB}{R\,dt} = -\frac{SS_1}{\rho l}\frac{dB}{dt},$$

where it is accepted that $R = \rho\frac{l}{S_1}$, is the area of section of a tissue through which the vortical current flows, l - length of an electro-conducting contour. The area through which the vortical current flows is a ring; part of which is shown in figure 2.3.4. Areas S and are perpendicular to each other.

The area of a ring is proportional to the thickness of a skin-layer $S_1 \sim \lambda$. Area S is in a plane; figure 2.3.4. It is also proportional to the thickness of a skin-layer $S \sim \lambda$.

Taking into account the oscillatory character of magnetic field induction $B = B_{\max} \cos \omega t$, we shall find:

$$I \sim \frac{\lambda^2}{\rho} \omega B_{\max} \sin \omega t = \frac{\lambda^2}{\rho} \omega B_{eff} , \tag{2.3.22}$$

where the oscillatory value of magnetic field induction is replaced with the effective value equivalent on energy.

Under Biot-Savart-Laplace law the magnetic field induction in a tissue is proportional to an external current:

$$B \sim i = \frac{E}{\rho} S_1 \sim \frac{E}{\rho} \lambda . \tag{2.3.23}$$

The equality (2.3.23) is correct for effective values of the magnetic field induction and of electric field strength.

Hence:

$$I \sim \frac{\lambda^2}{\rho} \omega B_{eff} \sim \frac{\lambda^3}{\rho^2} \omega E_{eff} . \tag{2.3.24}$$

The quantity of heat allocated in a unit of volume in a unit of time of the electro-conducting tissue (heat production) with the account (2.3.24) equals:

$$q = \frac{I^2 R}{V} = \frac{I^2 \rho l}{V S_1} \sim \frac{I^2 \rho}{\lambda^2} \sim \frac{\lambda^6 \omega^2 E_{eff}^2 \rho}{\rho^4 \lambda^2} \sim \frac{\lambda^4 \omega^2 E_{eff}^2}{\rho^3} . \tag{2.3.25}$$

At the deduction (2.3.25) it is taken into account that volume $V = S_1 L \sim \lambda$, where L is the length of the heated up body, figure 2.3.4.

Using $\lambda = \sqrt{\frac{2\rho}{\mu_a \omega}} \sim \sqrt{\frac{\rho}{\omega}}$, we find:

$$q \sim \frac{\lambda^4 \omega^2 E_{eff}^2}{\rho^3} \sim \frac{E_{eff}^2}{\rho} . \tag{2.3.26}$$

From the general physical reasons, with the absence of losses of electromagnetic field energy in the surrounding space, it follows that the factor before the right part in (2.3.26) is equal to unit $q = \frac{E_{eff}^2}{\rho}$.

The completed analysis shows that the absence of dependence heat production q in electro-conducting tissues from the frequency of electromagnetic field ω is connected to the occurrence in such tissues of a skin-layer. The allocation of heat in electro-conducting tissues occurs in the field of a skin-layer which, with increase in frequency, decreases.

Therefore, the increase in heat production along with an increase in frequency is compensated by a reduction of volume of a tissue where there is an allocation of heat and dependence heat production from a frequency not observed. One of the features of UHF - therapy consists in it.

2.3.2.2. UHF - Heating up of Dielectric Tissues

For the description of dielectric properties of substances we use the approach accepted in electrical engineering that wassuggested by Hippel [9]. This approach obviously is not connected to the application of Maxwell's equations, but it is effective enough.

Let's model a dielectric tissue with an equivalent electric circuit having a condenser and a resistor connected in parallel. The biological tissue in a condition of rest has active and capacitor resistance.

The current of conductivity (active component of a current) and current of displacement (capacitor component of a current) do not coincide on a phase. The capacitor component of a current outstrips the active, as shown in the vector diagram, at an angle of 90^0, as in figure 2.3.5.

The active component of the current and voltage coincide on a phase. It is necessary to note that the current and voltage are scalar values. Their vector representation is conditional.

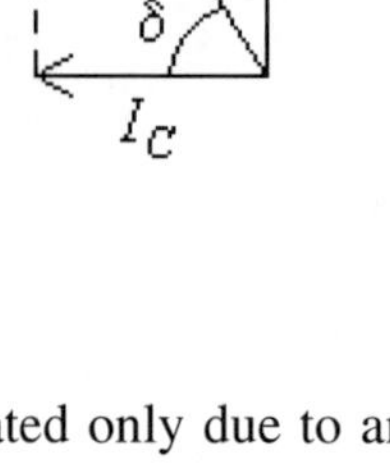

Figure 2.3.5.

As the heat in the tissues is allocated only due to an active component of a current I_R, the power allocated in the tissue is:

$$P = U_{eff} I_{eff} \cos\varphi = \frac{U}{\sqrt{2}} \frac{I}{\sqrt{2}} \cos\varphi = \frac{UI_R}{2} = \frac{UI_C \text{tg}\delta}{2}, \qquad (2.3.27)$$

where *I* and *U* are amplitude sizes full of a current and voltage, and - their effective values, φ - an angle (a difference of phases) between a current and voltage - an angle of dielectric losses, see (2.2.22). For example, the size of an angle of dielectric losses for a skeletal muscle of men in a range of frequencies UHF - therapy is equal to [6]. The capacitor component of the current equals:

$$I_C = \frac{U}{X_C} = U\omega C = U\omega\varepsilon\varepsilon_0 \frac{S_2}{L}, \qquad (2.3.28)$$

where is a capacitive reactance of the medium, - capacity of the condenser formed by electrodes, and *L* - the area and the linear size of the heated up body. In figure 2.3.4, the volume of the body is given as $V = S_2 L$.

Hence, substituting the effective value of voltage $U_{eff} = \frac{U}{\sqrt{2}}$, we shall find the thermal power allocated in the tissue:

$$P = \frac{U^2 \omega\varepsilon\varepsilon_0 S_2 \text{tg}\delta}{2L} = \frac{U_{eff}^2 \omega\varepsilon\varepsilon_0 S_2 \text{tg}\delta}{L} = \frac{(E_{eff} L)^2 \omega\varepsilon\varepsilon_0 S_2 \text{tg}\delta}{L} = E_{eff}^2 \omega\varepsilon\varepsilon_0 V \text{tg}\delta \qquad (2.3.29)$$

The quantity of heat allocated in a unit of volume in a unit of time of a dielectric tissue equals:

$$q = \frac{P}{V} = k_1 E_{eff}^2 \omega\varepsilon\varepsilon_0 \text{tg}\delta, \qquad (2.3.30)$$

where is the factor reflecting losses of energy on radiation in the environment (2.3.13). Thus, the greater the angle of dielectric losses δ, the more active the component of a current and, hence, the more energy that is allocated as heat in a dielectric tissue.

2.4. Use of Maxwell's Equations for the Analysis of Optical Active Substances

Substances capable of rotating a plane of polarization of light are referred to as being optically active. Optically active substances include: a solution of sugar, turpentine, quartz, etc. Optical activity is connected to the asymmetry of a structure of molecules. The molecule can rotate a plane of polarization if it has no center or plane of symmetry.

Asymmetric molecules usually have two versions: left-handed (rotating a counter-clockwise plane of polarization of light) or *L*-form and right-handed (rotating a clockwise plane of polarization of light) or *D*-form, figure 2.4.1. Molecules of these substances represent a mirror reflection of one another and are referred to as chiral (from the Greek – hand). Substances capable of rotating a plane of polarization of electromagnetic waves are also frequently referred to as chiral.

Figure 2.4.1.

2.4.1. The Phenomenological Theory of Optical Activity

In connection with that, the need for understanding optical activity has been apparent since its discovery in 1811 by the French scientist D. F. Arago, and representations about the electromagnetic nature of light were still in their infancy; the French physicist Fresnel in 1823 offered the phenomenological theory of optical activity. The essence of this theory will consist in the following:

A light vector (a vector of electric field strength in electromagnetic wave) in a plainly polarized light wave propagating along an axis *X*, it is possible to assume as the sum of two light vectors rotating in the different sides, as shown in figure 2.4.2.

If speeds of wave propagation polarized on a circle to the right and to the left are various, the medium has circular double-refraction.

Optical activity is the result of circular double-refraction.

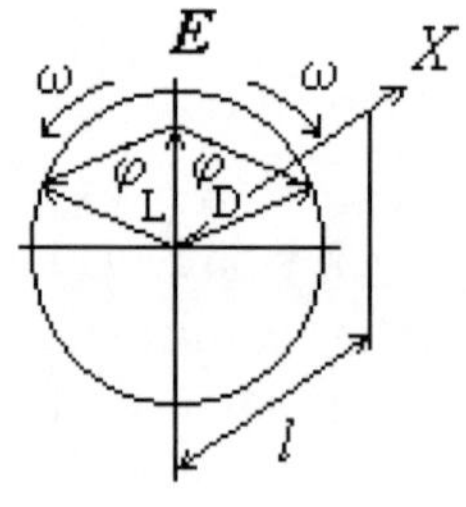

Figure 2.4.2.

Linear speeds of propagation rotating to the right and to the left light vectors along an axis *X* are accordingly equal to and V_L, and the angular speeds of their rotation ω are identical. Therefore, with a double angle of rotation φ of a plane of polarization, i.e. a plane of oscillations of a light vector $\boldsymbol{E}$, it is possible to find from a condition:

$$2\varphi = \varphi_D - \varphi_L = \omega(t_D - t_L) = 2\pi\nu\left(\frac{l}{V_D} - \frac{l}{V_L}\right) = \frac{2\pi d}{\lambda}(n_D - n_L)$$

Hence:

$$\varphi = \frac{\pi d}{\lambda}(n_D - n_L). \tag{2.4.1}$$

where and are the angles of rotation of the right and left rotating light vectors, and - times during which these vectors will pass one way l, ν - linear frequency of a light wave, λ - length of a light wave, and - parameters of refraction for the light waves polarized on a circle.

The reason for the occurrence of a double angle can be understood if we assume that the light vector with the left-circular polarization does not rotate, i.e. $\varphi_L = 0$. Then the bisector between rotating and not rotating the light vectors, located under an angle $\varphi = \frac{\varphi_D}{2}$, characterizes the rotation of a plane of polarization of a light vector $\boldsymbol{E}$.

If circular components of plainly polarized light propagate in a medium not only with different speed, but also are variously absorbed, light propagated through such medium will be elliptically polarized. This phenomenon refers to circular dichroism. The simultaneous existence of circular double-refraction and circular dichroism is referred to as the Cotton effect.

Taking into account the fact that the parameter of the refraction of substance is proportional to the concentration C of optically active substances in a solution, i.e. $(n_D - n_L) \sim C$, we find an angle of polarization plane in the rotation of monochromatic light of an optically active substance:

$$\varphi = \alpha C l,$$

where l is the thickness of a solution, α - the specific rotation which is numerically equal to an angle of polarization plane of rotation in a solution of optically active substance of unit concentration and unit thickness.

For a pure substance, for example, quartz, the angle of polarization of the plane of rotation is determined by the formula:

$$\varphi = \alpha l, \tag{2.4.2}$$

where α is also referred to as specific rotation.

For calibration of optical devices in which optical activity of substances is investigated, specific rotation usually measures the monochromatic light corresponding yellow D - lines in a spectrum of couples of sodium (length of a wave $\lambda = 589.3\ nm$) at $t = 20^0 C$.

Aside from specific rotation α there is also usually molecular rotation:

$$M = \frac{\mu}{100}\alpha, \tag{2.4.3}$$

where μ is the molecular mass of the substance.

For the analysis it is convenient to assume a parameter of refraction complex value, see (2.2.17):

$$n' = n - ik, \tag{2.4.4}$$

where n is a usual parameter of refraction, k - a parameter of absorption of substance.

Therefore, the angle of polarization plane of rotation (2.4.1) also can be written down in complex values:

$$\varphi' = \frac{\pi d}{\lambda}\left(n'_D - n'_L\right) = \frac{\pi d}{\lambda}\left(n_D - n_L\right) - i\frac{\pi d}{\lambda}\left(k_D - k_L\right), \tag{2.4.5}$$

where and are parameters of absorption for the right and left rotating light vectors.

With elliptic polarization the value refers to as an ellipticality of light:

$$\theta = \frac{\pi}{\lambda}\left(k_D - k_L\right). \tag{2.4.6}$$

Similarly, (2.4.3) also uses molecular ellipticality:

$$\theta_{mol} = \frac{\mu}{100}\theta. \tag{2.4.7}$$

Ellipticality and molecular ellipticality are characterized by the circular dichroism of the medium.

2.4.2. Electrodynamics of Optical Activity

Research of a chiral or optically active medium can be carried out on the basis of Maxwell's equations. However, as opposed to the problems solved in paragraph 2.2 and 2.3, it is necessary to use the modifications in comparison with (2.1.12) and (2.1.13) material equations. We shall make a statement in the system of units CGS.

For example, frequently the material equations look like [10]:

$$\boldsymbol{D} = \varepsilon \boldsymbol{E} \mp i\chi \boldsymbol{H}, \tag{2.4.8}$$
$$\boldsymbol{B} = \mu \boldsymbol{H} \pm i\chi \boldsymbol{E}, \tag{2.4.9}$$

where is the parameter of chirality, i - the imaginary unit, the top signs correspond to the right rotating chiral medium, and the bottom signs to the left rotating medium.

We can transform the material equations to a more standard form. For this purpose, using Fresnel's idea, we shall present vectors of electric field strength $\boldsymbol{E}$ and magnetic field strength $\boldsymbol{H}$ in the electromagnetic wave as the sum of two vectors rotating in the different sides, as in figure 2.4.2:

$$\boldsymbol{E} = \boldsymbol{E}_D + \boldsymbol{E}_L = -i\sqrt{\frac{\mu}{\varepsilon}}(\boldsymbol{H}_D + \boldsymbol{H}_L)$$

and

$$\boldsymbol{H} = \boldsymbol{H}_D + \boldsymbol{H}_L = i\sqrt{\frac{\varepsilon}{\mu}}(\boldsymbol{E}_D + \boldsymbol{E}_L),$$

where and are auxiliary vectors of electric and magnetic fields' strength with the right circular (RCP), and and with left-circular (LCP) polarization. Connection between magnetic and electric fields' strength in electromagnetic wave (2.2.23) is taken into account also. An imaginary unit is the consequence of using complex approximation of the material equations (2.4.8) and (2.4.9).

We transform the material equation (2.4.8) to vectors and $\boldsymbol{E}_L$:

$$\boldsymbol{D} = \boldsymbol{D}_D + \boldsymbol{D}_L = \varepsilon(\boldsymbol{E}_D + \boldsymbol{E}_L) \pm \chi\sqrt{\frac{\varepsilon}{\mu}}(\boldsymbol{E}_D + \boldsymbol{E}_L),$$

$$\boldsymbol{D}_{D,L} = \varepsilon \boldsymbol{E}_{D,L} \pm \chi\sqrt{\frac{\varepsilon}{\mu}}\boldsymbol{E}_{D,L} = \varepsilon\left(1 \pm \frac{\chi}{n}\right)\boldsymbol{E}_{D,L}, \tag{2.4.10}$$

In the material equation (2.4.9) we shall transit to vectors and $\boldsymbol{H}_L$:

$$\boldsymbol{B} = \boldsymbol{B}_D + \boldsymbol{B}_L = \mu(\boldsymbol{H}_D + \boldsymbol{H}_L) \pm \chi\sqrt{\frac{\mu}{\varepsilon}}(\boldsymbol{H}_D + \boldsymbol{H}_L),$$

$$\boldsymbol{B}_{D,L} = \mu \boldsymbol{H}_{D,L} \pm \chi\sqrt{\frac{\mu}{\varepsilon}}\boldsymbol{H}_{D,L} = \mu\left(1 \pm \frac{\chi}{n}\right)\boldsymbol{H}_{D,L}, \tag{2.4.11}$$

Hence, the material equations can be written down as:

$$\boldsymbol{D}_{D,L} = \varepsilon_{D,L}\boldsymbol{E}_{D,L} \text{ and } \boldsymbol{B}_{D,L} = \mu_{D,L}\boldsymbol{H}_{D,L}, \tag{2.4.12}$$

where

$$\varepsilon_{D,L} = \varepsilon\left(1 \pm \frac{\chi}{n}\right) \text{ and } \mu_{D,L} = \mu\left(1 \pm \frac{\chi}{n}\right). \tag{2.4.13}$$

As before, the top signs are chiral and correspond to the right rotating medium, and the bottom signs to the left rotating medium.

Thus, waves RCP and LCP in a chiral medium have various parameters of refraction:

$$n_{D,L} = \sqrt{\varepsilon_{D,L}\mu_{D,L}} = n \pm \chi, \tag{2.4.14}$$

from which difference of phase speeds of these waves follows. Hence, at the propagation in chiral medium the polarization plane of electromagnetic wave rotates on some angle. Let's find an angle of polarization plane that rotates at the passage of a wave through the chiral layer. Substituting expressions for and from (2.4.14) in Fresnel's formula (2.4.1), we obtain:

$$\varphi = \frac{\pi l}{\lambda}(n_D - n_L) = \frac{2\pi\chi l}{\lambda} = p\chi l = \alpha l, \tag{2.4.15}$$

where, see (2.4.2), specific rotation $\alpha = p\chi$, and value is the wave number of the electromagnetic wave propagating in the chiral medium.

Thus, the physical sense of the parameter of chirality consists in its equality to the specific rotation divided into the wave number:

$$\chi = \frac{\alpha}{p}. \tag{2.4.16}$$

Hence, there is no necessity to use two characteristics of a chiral medium, and it is possible to use either parameter of chirality or specific rotation.

Let's consider the process of propagation of a plane electromagnetic wave in a chiral medium. Maxwell's equations for an electromagnetic field in this medium, in the absence of free charges and currents, we shall write down as:

$$\text{rot}\boldsymbol{H} = \frac{1}{c}\frac{\partial \boldsymbol{D}}{\partial t}, \tag{2.4.17}$$

$$\text{rot}\boldsymbol{E} = -\frac{1}{c}\frac{\partial \boldsymbol{B}}{\partial t}, \tag{2.4.18}$$

$$\text{div}\boldsymbol{H} = 0, \tag{2.4.19}$$

$$\text{div}\boldsymbol{E} = 0, \tag{2.4.20}$$

where c is the speed of light in vacuum.

Let's assume, for simplification of the problem, that absorption of an electromagnetic wave in the medium does not occur; i.e. the medium is characterized only by double-refraction. In this case, for solving equations (2.4.17) - (2.4.20) we shall find the harmonious waves propagating in a direction of an axis X *as*:

$$\boldsymbol{B} = \boldsymbol{B}_0 \exp i(\omega t - pX), \tag{2.4.21}$$

$$\boldsymbol{D} = \boldsymbol{D}_0 \exp i(\omega t - pX), \tag{2.4.22}$$

where and are amplitudes of an induction of magnetic and electric waves in the medium, ω - cyclic frequency of waves. For these waves formulas and are correct

Hence, Maxwell's equations (2.4.17) and (2.4.18) can be written down as:

$$\text{rot}\boldsymbol{H} = i\frac{\omega}{c}\boldsymbol{D} = \frac{p}{n}(i\varepsilon\boldsymbol{E} \pm \chi\boldsymbol{H}), \tag{2.4.23}$$

$$\text{rot}\boldsymbol{E} = -i\frac{\omega}{c}\boldsymbol{B} = \frac{p}{n}(-i\mu\boldsymbol{H} \pm \chi\boldsymbol{E}), \tag{2.4.24}$$

where wave number is $p = \dfrac{n\omega}{c}$.

Let's apply to the equations (2.4.23) and (2.4.24) the operation of a rotor:

$$\text{graddiv}\,\boldsymbol{H} - \Delta\boldsymbol{H} = \frac{p^2}{n^2}\left[\left(\varepsilon\mu + \chi^2\right)\boldsymbol{H} \pm 2i\varepsilon\chi\boldsymbol{E}\right],$$

$$\text{graddiv}\,\boldsymbol{E} - \Delta\boldsymbol{E} = \frac{p^2}{n^2}\left[\left(\varepsilon\mu + \chi^2\right)\boldsymbol{E} \mp 2i\mu\chi\boldsymbol{H}\right].$$

Taking into account Maxwell's equations (2.4.19) and (2.4.20), we shall receive two connected differential equations of the second order concerning vectors $\boldsymbol{H}$ and $\boldsymbol{E}$:

$$\Delta \boldsymbol{H} + \frac{p^2}{n^2}\left[\left(n^2+\chi^2\right)\boldsymbol{H} \pm 2i\varepsilon\chi\boldsymbol{E}\right] = 0, \tag{2.4.25}$$

$$\Delta \boldsymbol{E} + \frac{p^2}{n^2}\left[\left(n^2+\chi^2\right)\boldsymbol{E} \mp 2i\mu\chi\boldsymbol{H}\right] = 0, \tag{2.4.26}$$

where the formula for a parameter of refraction is used.

We transform the equations (2.4.25) and (2.4.26) to vectors and $\boldsymbol{E}_L$:

$$\Delta(\boldsymbol{E}_D + \boldsymbol{E}_L) + \frac{p^2}{n^2}\left[\left(n^2+\chi^2\right)(\boldsymbol{E}_D + \boldsymbol{E}_L) \pm 2n\chi(\boldsymbol{E}_D + \boldsymbol{E}_L)\right] = 0 \cdot \tag{2.4.27}$$

The equation (2.4.26) results in the same equation. The equation (2.4.27), under the condition of independence of vectors and $\boldsymbol{E}_L$, is possible to break into two equations:

$$\Delta \boldsymbol{E}_D + p^2\left(1+\frac{\chi}{n}\right)^2 \boldsymbol{E}_D = 0, \tag{2.4.28}$$

$$\Delta \boldsymbol{E}_L + p^2\left(1-\frac{\chi}{n}\right)^2 \boldsymbol{E}_L = 0, \tag{2.4.29}$$

where it is taken into account that the top signs are chiral and correspond to the right rotating medium, and the bottom signs to the left rotating medium.

The equations (2.4.28) and (2.4.29) describe the propagation of two electromagnetic waves with RCP and LCP and with various wave numbers:

$$p_{D,L} = p\left(1 \pm \frac{\chi}{n}\right). \tag{2.4.30}$$

There are also other forms of writing the material equations of chiral mediums. For example, investigating a chiral medium as a substance with a small spatial dispersion in [4, 11], it is possible to use the tensor form of the material equations:

$$\boldsymbol{D} = \varepsilon_{ij}\boldsymbol{E} \text{ and } \boldsymbol{B} = \mu_{ij}\boldsymbol{H}, \tag{2.4.31}$$

where

$$\varepsilon_{ij} = \frac{1}{p\mu}\begin{pmatrix} p\varepsilon\mu & \alpha_3 & -\alpha_2 \\ -\alpha_3 & p\varepsilon\mu & \alpha_1 \\ \alpha_2 & -\alpha_1 & p\varepsilon\mu \end{pmatrix} \text{ and } \mu_{ij} = \frac{1}{p\varepsilon}\begin{pmatrix} p\varepsilon\mu & \alpha_3 & -\alpha_2 \\ -\alpha_3 & p\varepsilon\mu & \alpha_1 \\ \alpha_2 & -\alpha_1 & p\varepsilon\mu \end{pmatrix}, \tag{2.4.32}$$

or, taking into account $n^2 = \varepsilon\mu$, we have:

$$\varepsilon_{ij} = \frac{1}{p\mu}\begin{pmatrix} pn^2 & \alpha_3 & -\alpha_2 \\ -\alpha_3 & pn^2 & \alpha_1 \\ \alpha_2 & -\alpha_1 & pn^2 \end{pmatrix} \text{ and } \mu_{ij} = \frac{1}{p\varepsilon}\begin{pmatrix} pn^2 & \alpha_3 & -\alpha_2 \\ -\alpha_3 & pn^2 & \alpha_1 \\ \alpha_2 & -\alpha_1 & pn^2 \end{pmatrix}. \tag{2.4.33}$$

In the formula (2.4.33), $\alpha_i = p_i\chi$, and are projections to axes of the coordinates of a wave vector in a chiral medium.

Apparently, from tensors dielectric and magnetic permeability (2.4.33), there are specific rotations around the axes $i = x, y, z$ in them. Specific rotation is not included in diagonal elements slanting-symmetric tensors and μ_{ij} . As is known, non-diagonal elements slanting-symmetric tensors describe the effects of rotation since rotation is an asymmetrical process. Angles of polarization plane rotation around different axes can be various. Owing to these reasons, it is possible to assume that the material equations (2.4.31) are fuller than (2.4.8), (2.4.9), reflecting the properties of the chiral medium.

2.4.3. The Molecular Theory of Optical Activity

Let's consider some electrodynamical aspects of the molecular theory of optical activity.

Let's assume that the reaction of separate spherical optical active molecules on an external electromagnetic field looks like:

$$\boldsymbol{p} = \alpha\boldsymbol{E}^{/} \mp \frac{\beta}{c}\frac{\partial\boldsymbol{H}}{\partial t}, \tag{2.4.34}$$

where is induced under the action of electromagnetic wave of the electric dipole moment of a molecule, α in is this case the polarizability of a molecule, see (1.2.7), $\boldsymbol{E}$ and $\boldsymbol{H}$ are vectors of electric and magnetic fields' strength in the electromagnetic wave propagating in substance, [4] - of an external electric field strength, is a vector of polarization of the medium, c - speed of light in vacuum. Parameter β characterizes the optical activity and is referred to as the rotating ability of the medium. The top signs chiral corresponds to the right rotating medium, and the bottom signs to the left rotating medium.

Using (1.2.8) and (2.4.34), we find the full electric moment of the unit of volume of the medium:

$$\boldsymbol{P} = n_0 \boldsymbol{p} = n_0 \alpha\left(\boldsymbol{E} + \frac{4\pi}{3}\boldsymbol{P} \right) \mp n_0 \frac{\beta}{c}\frac{\partial \boldsymbol{H}}{\partial t}, \tag{2.4.35}$$

Uniting addends in (2.4.35), we shall allocate a vector of polarization $\boldsymbol{P}$:

$$\boldsymbol{P} = \frac{n_0 \alpha}{\left(1 - n_0 \alpha \frac{4\pi}{3}\right)} \boldsymbol{E} \mp \frac{n_0}{\left(1 - n_0 \alpha \frac{4\pi}{3}\right)} \frac{\beta}{c}\frac{\partial \boldsymbol{H}}{\partial t}. \tag{2.4.36}$$

Let's find an electric field induction in the optical active medium:

$$\boldsymbol{D} = \boldsymbol{E} + 4\pi \boldsymbol{P} = \left[1 + \frac{4\pi n_0 \alpha}{\left(1 - n_0 \alpha \frac{4\pi}{3}\right)} \right] \boldsymbol{E} \mp \frac{4\pi n_0}{\left(1 - n_0 \alpha \frac{4\pi}{3}\right)} \frac{\beta}{c}\frac{\partial \boldsymbol{H}}{\partial t} =, \tag{2.4.37}$$

$$= \varepsilon \boldsymbol{E} \mp \frac{4\pi n_0}{\left(1 - n_0 \alpha \frac{4\pi}{3}\right)} \frac{\beta}{c}\frac{\partial \boldsymbol{H}}{\partial t}$$

where is the relative dielectric penetrance of the medium, is the relative dielectric susceptibility of the substance (the index 0 is used for the difference of a relative dielectric susceptibility from a parameter of chirality). Further, it is simple to have the formula of Clausius - Mossotti (1.2.13).

The formula (2.4.37) for an electric induction can be written down as:

$$\boldsymbol{D} = \varepsilon \boldsymbol{E} \mp \frac{(\varepsilon - 1)}{\alpha} \frac{\beta}{c} \frac{\partial \boldsymbol{H}}{\partial t} \tag{2.4.38}$$

Using Maxwell's equation (2.4.18) $\mathrm{rot}\boldsymbol{E} = -\frac{\mu}{c}\frac{\partial \boldsymbol{H}}{\partial t}$, we find:

$$\boldsymbol{D} = \varepsilon \boldsymbol{E} \pm \frac{(\varepsilon - 1)}{\alpha} \frac{\beta}{\mu} \mathrm{rot}\boldsymbol{E}. \tag{2.4.39}$$

Furthermore, we use the Maxwell's equation $\mathrm{rot}\boldsymbol{E} = -i\frac{p}{n}\boldsymbol{B}$, where p - wave number, for harmoniously oscillating the electromagnetic field. Taking into account $\boldsymbol{B} = \mu \boldsymbol{H}$, we find:

$$\boldsymbol{D}=\varepsilon\boldsymbol{E}\mp ip\frac{(\varepsilon-1)}{\alpha}\frac{\beta}{n\mu}\boldsymbol{B}=\varepsilon\boldsymbol{E}\mp ip\frac{(\varepsilon-1)}{\alpha}\frac{\beta}{n}\boldsymbol{H}\cdot \tag{2.4.40}$$

Thus, taking into account the material equation (2.4.8) for the chiral medium $\boldsymbol{D}=\varepsilon\boldsymbol{E}\mp i\chi\boldsymbol{H}$, we find that the parameter of chirality χ is connected to the rotating ability of medium β under the formula:

$$\chi=p\frac{(\varepsilon-1)}{\alpha n}\beta. \tag{2.4.41}$$

Using the connection of the parameter chirality χ with an angle of polarization plane rotation of an optically active substance (2.4.15), we find:

$$\varphi=p\chi l=p^2\frac{(\varepsilon-1)}{\alpha n}\beta l\cdot \tag{2.4.42}$$

where l is the path taken by an electromagnetic wave in the optically active medium. It should be noted that the angle of rotation of the polarization plane in the chiral medium according to (2.4.42) is inversely proportional to a square of length of a light wave $\varphi\sim p^2=\dfrac{4\pi^2}{\lambda^2}$ which corresponds to the Biot law.

2.5. Nonlinear Problems of Interaction of Electromagnetic Radiation with Substance on the Basis of Maxwell's Equations

The formation of electromagnetic impulse in substance is determined by the laws of interaction of electromagnetic radiation with atoms of substance. The form of an impulse is determined by the processes of polarization of atoms. But in connection with the nonlinearity of displacement of the centers of positive and negative charges in an atom in response to the action of the electromagnetic field, the problem frequently has an essentially nonlinear character. Thus, there are features of the process of formation of an impulse depending on its power.

2.5.1. Formation of Low Power Electric Impulse in Dielectric

Let's consider the one-dimensional problem of interaction of an electromagnetic field with a dielectric substance in which there is some numerical concentration of atoms-oscillators. For the purposes of the analysis we shall assume the atom is one-electronic.

The system of Maxwell's equations for electromagnetic fields in dielectric substances, in which there are no free charges or currents, looks like:

$$\mathrm{rot}\boldsymbol{H} = \frac{\partial \boldsymbol{D}}{\partial t} = \varepsilon_0 \frac{\partial \boldsymbol{E}}{\partial t} + \frac{\partial \boldsymbol{P}}{\partial t}, \quad (2.5.1)$$

$$\mathrm{rot}\boldsymbol{E} = -\mu\mu_0 \frac{\partial \boldsymbol{H}}{\partial t}, \quad (2.5.2)$$

$$\mathrm{div}\boldsymbol{E} = 0, \quad (2.5.3)$$

$$\mathrm{div}\boldsymbol{H} = 0, \quad (2.5.4)$$

where the interrelation between the electric induction and a vector of polarization is used as (1.2.14). Let's find a rotor from the equation (2.5.2):

$$\mathrm{rotrot}\boldsymbol{E} = -\Delta\boldsymbol{E} = -\mu\mu_0 \frac{\partial \mathrm{rot}\boldsymbol{H}}{\partial t} = -\mu\mu_0\varepsilon_0 \frac{\partial^2 \boldsymbol{E}}{\partial t^2} - \mu\mu_0 \frac{\partial^2 \boldsymbol{P}}{\partial t^2} \quad (2.5.5)$$

where the equation (2.5.3) is used.

Taking into account the speed of a wave in substance $V = \dfrac{1}{\sqrt{\varepsilon\varepsilon_0\mu\mu_0}}$, we shall find:

$$\Delta\boldsymbol{E} - \frac{1}{\varepsilon V^2}\frac{\partial^2 \boldsymbol{E}}{\partial t^2} = \frac{1}{\varepsilon_0 \varepsilon V^2}\frac{\partial^2 \boldsymbol{P}}{\partial t^2}. \quad (2.5.6)$$

The equation of wave propagation in the direction of an axis *X* for an electric component of field *E* can be written down as:

$$\frac{\partial^2 E}{\partial X^2} - \frac{1}{\varepsilon V^2}\frac{\partial^2 E}{\partial t^2} = \frac{n_0}{\varepsilon_0 \varepsilon V^2}\frac{\partial^2 p}{\partial t^2}, \quad (2.5.7)$$

where it is accepted that $P = n_0 p$, p is the module of a vector of polarization for one atom - the module of the electric dipole moment.

According to Newton's second law for the oscillator (an electronic orbit), in the absence of attenuation, (see figure 1.2.1) we have:

$$\frac{\partial^2 p}{\partial t^2} + \omega_0^2 p = \frac{e^2}{m} NE, \quad (2.5.8)$$

$$\omega_*^2 = \frac{2n_0 e^2}{\varepsilon_0 \varepsilon m}. \tag{2.5.17}$$

As a result, we shall find:

$$2i\frac{\omega_0}{\varepsilon}\frac{\partial \Phi}{\partial t} + V^2\frac{\partial^2 \Phi}{\partial X^2} = -\omega_*^2\left(\frac{\omega_0{}^2}{\omega_*^2} + \ln\left|\frac{\Phi}{\Phi_0^*}\right|\right)\Phi \cdot \tag{2.5.18}$$

Designating and $|\Phi_0| = |\gamma\Phi_0^*|$, we shall obtain:

$$2i\frac{\omega_0}{\varepsilon}\frac{\partial \Phi}{\partial t} + V^2\frac{\partial^2 \Phi}{\partial X^2} = -\omega_*^2\left(\ln\left|\frac{\Phi}{\Phi_0}\right|\right)\Phi \cdot \tag{2.5.19}$$

Let's divide the received equation of ω_0, and we shall designate the wave number $k = \frac{\omega_0}{V}$, and also and $\beta = \frac{2}{\varepsilon}$; as a result, we shall find:

$$i\beta\frac{\partial \Phi}{\partial t} + \frac{V}{k}\frac{\partial^2 \Phi}{\partial X^2} = -\omega\left(\ln\left|\frac{\Phi}{\Phi_0}\right|\right)\Phi \cdot \tag{2.5.20}$$

The equation (2.5.20) belongs to a class of nonlinear wave of Schrödinger's equations with logarithmic nonlinearity.

Through the solving of the equation (2.5.20) we shall find the following [14]:

$$\Phi = |\Phi_0| f(kX - \omega t)\exp[i(rX - \delta t)]. \tag{2.5.21}$$

After substitution (2.5.21) in (2.5.20) and of some transformations, we shall obtain:

$$i\beta\left[\frac{\partial f}{\partial t} - i\delta f\right] + \frac{V}{k}\left[\frac{\partial^2 f}{\partial X^2} + 2\frac{\partial f}{\partial X}ir - fr^2\right] = -\omega(\ln f)f\,.$$

Let's assume $f = C\exp\left[-\frac{(kX - \omega t)^2}{4}\right]$, where is an unknown constant. Substituting function *f* in the previous equation, we shall find:

$$\left[\frac{1}{2}(kX-\omega t)\omega i\beta+\delta\beta\right]+\frac{V}{k}\left[\left(\frac{1}{2}(kX-\omega t)k\right)^2-\frac{k^2}{2}-(kX-\omega t)kir-r^2\right]=.$$
$$=-\omega\left(\ln C-\frac{(kX-\omega t)^2}{4}\right)$$

Equating factors at identical degrees $(kX-\omega t)$, we shall find:

$$\beta=\frac{2Vr}{\omega}=2\frac{r}{k};\ Vk=\omega=\omega_0;\ C=\exp\left(\frac{r^2}{k^2}-\frac{2\delta r}{\omega k}+\frac{1}{2}\right).$$

Hence, the solving of the equation (2.5.20) looks like:

$$\Phi=|\Phi_0|\exp\left(\frac{r^2}{k^2}-2\frac{\delta r}{\omega k}+\frac{1}{2}\right)\cdot\exp\left[-\frac{(kX-\omega t)^2}{4}\right]\exp[i(rX-\delta t)] \qquad (2.5.22)$$

The solution (2.5.22) describes the solitary wave as having an envelope, which follows Gauss' law.

Thus, the following ratio $\beta=2\frac{r}{k}=\frac{2}{\varepsilon}$, i.e. $\frac{r}{k}=\frac{1}{\varepsilon}$, where r is the wave number and δ - bearing frequency of a filling wave, - the group speed of a wave package are carried out.

Taking into account $\omega=\frac{\omega_*^2}{\omega_0}$, we find $\omega=\omega_*=\omega_0=\sqrt{\frac{2n_0e^2}{\varepsilon_0\varepsilon m}}$.

The received ratio shows that the influence of dielectric on some electromagnetic impulse of very small duration (within infinitesimal limits) in dielectric substances generate an impulse, the frequency of which will be determined by the parameters of the dielectric.

According to the transformation (2.5.10), we shall transit in (2.5.22) to an electric field strength E. We use $|\Phi_0|=|\gamma\Phi_0^*|=\frac{E_0}{\mathrm{e}}$, since and according to (2.5.15). Hence:

$$E=E_0\exp\left(\frac{r^2}{k^2}-2\frac{\delta r}{\omega k}-\frac{1}{2}\right)\cdot\exp\left[-\frac{(kX-\omega t)^2}{4}\right]\exp[i(rX-(\delta+\omega)t)]\cdot$$

Usually, the frequency of a filling wave is much greater than that of an envelope wave $\delta\gg\omega_0=\omega$. Hence:

$$E=E_0\exp\left(\frac{r^2}{k^2}-2\frac{\delta r}{\omega k}-\frac{1}{2}\right)\cdot\exp\left[-\frac{(kX-\omega t)^2}{4}\right]\exp[i(rX-\delta t)] \qquad (2.5.23)$$

Before proceeding to a numerical illustrative calculation, it is necessary to make some comment on the result.

At the deduction of equation (2.5.20) and, accordingly, the solution (2.5.23), we used two physical models: the relativistic, absolutely correct model of Maxwell for propagation of the electromagnetic wave in the substance and the mechanical model of the atom with which this wave is to interact. Also, the model of the atom is very far from perfection. First, in it there are no quantum representations; it not relativistic, besides, the magnetic component of electromagnetic wave does not interact which considerably reduces the reliability of the examined process.

But another aspect is more important. The propagation of the electromagnetic impulse is a two-high-speed process. There is a group speed of propagation of a wave package. According to the solution (2.5.23), this speed is $V=\dfrac{\omega}{k}$. However, the filling wave is propagated with a phase speed $V_f=\dfrac{\delta}{r}$. And it is usually $V_f >> V$. In an examined picture of the interaction of an electromagnetic wave and atom there is only one speed in Maxwell's equations - the speed of propagation of the electromagnetic wave in the substance. The second speed could appear at the account of a phase delay of oscillations of the electron shell of the atom (1.2.3) concerning the oscillations of the electric component of wave. In using a model of the atom, such a phase delay is absent, since there is no attenuation of oscillations of the electron shell; i.e., finally, there is no reradiated reaction wave on the atom. In consideration of the aforementioned, we cannot take into account some ratios from the model; in particular, the interrelation of a filling wave and the envelope wave impulse in the calculation. The simplest kind of solution (2.5.23) has in the case of group speed of a wave package $V \approx c$, where c is the speed of light in a vacuum. It is approximately correct in a rarefied dielectric - air.

Let's carry out some numerical estimations. Modern lasers generate impulses with a duration of [12, 15]. This corresponds to a range of cyclic frequencies $\omega=\dfrac{2\pi}{T}=6.28\cdot\left(10^9 \div 10^{12}\right)s^{-1}$. Taking into account $\omega_* = \omega$, and also (2.5.17), we shall find the numerical concentration of the atoms $n_0=6{,}2\cdot\left(10^{15}\div 10^{21}\right)m^{-3}$.

Taking into account that in air under normal conditions the numerical concentration of molecules is $2.7\cdot 10^{25}\ m^{-3}$, the assumption about the rarefied dielectric medium is correct.

For calculation purposes, we use the frequency of an envelope wave. This corresponds to the wave package length $\lambda=\dfrac{2\pi V}{\omega}=3\cdot 10^{-4}\ m$, i.e. the top border of a radio-frequency range. Besides, we shall find the wave number of an envelope wave package $k=\dfrac{\omega}{V}=2.1\cdot 10^4 m^{-1}$. The period of a filling wave is $T_f=10^{-13}s$, which corresponds to the bearing cyclic frequency $\delta=\dfrac{2\pi}{T_f}=6.28\cdot 10^{13}\ s^{-1}$.

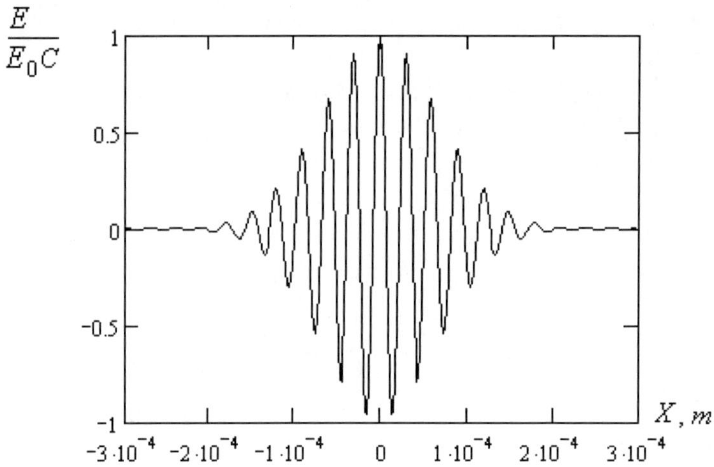

Figure 2.5.1.

The bearing cyclic frequency of a filling wave is not connected by any ratio to dielectric parameters. It is completely determined by the impulse frequency falling on the dielectric. The wave number of the filling wave under the condition of (which is correct in the rarefied air) can be found by using the formula $r=\dfrac{\delta}{V_f}=2.1\cdot 10^5 m^{-1}$. Using a designation $C=\exp\left(\dfrac{r^2}{k^2}-2\dfrac{\delta r}{\omega k}-\dfrac{1}{2}\right)$, we shall write down (2.5.23) as:

$$E=E_0 C\cdot\exp\left[-\frac{(kX-\omega t)^2}{4}\right]\exp[i(rX-\delta t)]\cdot \tag{2.5.24}$$

In figure 2.5.1 the graph of dependence of relative electric field strength in an impulse $\dfrac{E}{E_0 C}(X)$, plotted for the sizes of constant parameters under the formula (2.5.24) in the conditional moment of time $t=0$, is shown. The solution (2.5.22) has a soliton character. Unfortunately, a two-soliton solution of the nonlinear Schrödinger equation with logarithmic nonlinearity (2.5.20) is not known. However, in [16] the interaction of the solitary waves described by the equation (2.5.20) has been numerically analyzed. Thus, it is shown that they behave as solitons.

2.5.2. Formation of Electromagnetic High-Power Impulse in Dielectric: A Self-Induced Transparency

Let's consider the basic phenomenological laws describing the formation in dielectric electric and magnetic components of a high intensity impulse.

The description of the propagation of an impulse in the medium with dissipation of its energy exclusively represents a challenge. But in many practically important cases, the loss of energy of an impulse in the medium can be neglected. From this point of view, the simplest is the description of an impulse at self-induced transparency (SIT).

The self-induced transparency has a simple mechanical analogy [13]. Let's imagine that there is a number of identical ideal pendulums suspended in a line above a plane. The massive spheres are rolling on a plane and strike the first pendulum. If the energy of a sphere is rather insignificant, the sphere will give a part of its energy to the first pendulum, then to the second, etc.; yet the sphere does not use all of the impulse. But if the energy of a sphere is great, the first pendulum will make a full turn of the suspension axis and will strike the sphere from behind. Therefore, the sphere will not lose energy. The same thing will take place and at impact about the second pendulum, etc. Thus, the sphere will roll without losing energy.

The phenomenon of SIT can arise in rarefied gas ($n_0 < 10^{18}\ atoms/cm^3$) for short laser impulses ($t < 10^{-9}\ s$) at the energy of a sufficient impulse for translation in the excited condition of all atoms in the area of influence of the impulse [12]. In this case there is no return scattering of electromagnetic radiation, dissipative properties of the system "impulse - medium" disappear and it is transformed in the conservative state. The electromagnetic impulse becomes stationary, in a soliton state. The symmetry and stability of an impulse considerably simplify its mathematical description.

Until now, the mathematical description has been similar to an electromagnetic soliton, on the basis of using the half-classical system of Maxwell-Bloch equations [13], which, in our opinion, is in unsatisfactory condition. The Maxwell-Bloch equations were written in 1946, long before the creation of lasers and the discovery of the SIT phenomenon in 1965.

The physical bases for these equations, except Maxwell's equations, are; firstly, the second law of Newton for the atomic electron and secondly, the proportionality of the average by the number of atoms filling N in the field of impulse influence to the volumetric density of electromagnetic wave energy W, i.e. $N \sim W$. The value N provides the measure of inversion in the system of atom-radiators by excited atoms [12].

The procedure of electromagnetic impulse description, including wave equation usually passes through the stage of field strength separation to the low envelope amplitude and the filling wave. Such a course is typical for processes underSchrödinger's nonlinear equations. However, the description of envelope amplitude, described on the basis of the Maxwell-Bloch theory, is based on Sine-Gordon's equation [13], which does not contain a filling wave in its solution. Such a course is unlimited and internally contradicting. The reason for the limitation of the Maxwell-Bloch equation usage for the SIT description will be analyzed further. We have to assume that the consideration of the SIT process should be done on the basis of consecutive procedure of the nonlinear Schrödinger equation. However, the prevalent nonlinear Schrödinger equation with cube nonlinearity, which can produce the solitary wave with the filling, is inappropriate for the SIT description. The reason is that the soliton solving of the nonlinear Schrödinger equation with cube nonlinearity is related to the impulse, in which the phase velocity of filling wave is less than the velocity of the impulse itself. For the SIT impulse the inversed correlation of velocities is typical [17].

Let's consider the one-dimensional task of interaction between the electric and magnetic component of electromagnetic field impulse and the dielectric substance, which possesses a certain numerical concentration of center-symmetrical atoms – oscillators.

As in paragraph 2.5.1, we suggest that the atom should be one-electronic. It is also agreed that the micro current or free charge is not present in the medium. The peculiarities of interaction between the magnetic aspect of the impulse and the atoms will be considered later.

Though all formal analysis will be carried out within the framework of classical electrodynamics, for a more adequate description of the processes we will address some quantum representations. We accept that the interaction of quantums of electromagnetic radiation with atomic electrons will take place; thus, quantums are absorbed by the electrons. By gaining the energy of the quantums, the electrons shift to advanced energy levels. Further, by means of a resonate shift of electrons back, the quantums' radiation appears to shift forward. The considered medium lacks a non-radiating shift of electrons, i.e. the quantum energy is not transferred to the nucleus.

Thus, the absorption of electromagnetic radiation in the meaning of its energy dissipation in the substance, owing to SIT, is disregarded. There is only the atomic reradiation of quantums. Thus, the forefront of the impulse gives the energy atomic electrons of the medium which reserve it, and then the energy completely comes back in the impulse, forming back to the front.

Probabilities of absorption and radiation quantums by electrons in units of time with plenty of quantums in impulse can be assumed, according to Einstein, to be approximately identical [18].

For separate interaction of quantum of a monochromatic flux of radiation with electron this probability is also proportional to a cube of constant thin structure [19].

Let's consider a random value - number of interactions of quantums with atomic electrons in impulse. According to the Poisson law of distribution, the probability that no quantum of atomic electrons will be absorbed (no interaction will take place), at rather low probability of separate interaction, is equal to an exponent from a mathematical expectation of a random variable - an average of interactions λ of quantums and electrons in impulse, taken with the minus $p = \exp(-\lambda)$. Therefore, it is possible to assume that the intensity of unabsorbed impulse energy on its forefront by atomic electrons of medium is determined to be exponential by Bouguer's law (in German tradition - Beer law) (2.3.2).

We assume that the intensity of energy return by the atomic electrons in impulse on its reverse can also be described approximately by the Bouguer law with a negative parameter of absorption [20]. The probability of the non-absorption of electromagnetic radiation is $p = \dfrac{I}{I_0}$, where is the intensity of radiation falling on dielectric, I – is not absorbed (passed) intensity.

Under the exponent in Bouguer's law (2.3.2), there is a mathematical expectation of the random variable, submitting to Poisson distribution - an average of atoms interacting with electromagnetic radiation in the field of influence of impulse $\lambda = MN$, see (2.5.14). Hence, or $I = I_0 \exp(-\alpha l)$, where α is the parameter of interaction of electromagnetic wave with substance, l – length of interaction layer, - intensity falling on dielectric waves.

Submitting to the deduction of Bouguer's law as $p = \exp(-\lambda)$, apparently, is elementary.

Using Bouguer's law, we shall consider the powerful electromagnetic impulse propagating in dielectric.

Taking into account that the wave intensity is , we shall have:

$$\begin{pmatrix} |E| \\ |H| \end{pmatrix} = \begin{pmatrix} |E_0| \\ |H_0| \end{pmatrix} \exp\left(-\frac{\alpha}{2} l\right), \tag{2.5.25}$$

where $|E_0|$, – shows the amplitudes of electric and magnetic fields' strength of the impulse on the longitudinal coordinate $X = 0$, - length of an impulse path in dielectric.

In formula (2.5.25) and further the upper values in round brackets are concerned with the electric field, and lower – to the magnetic field of impulse.

Similarly to ratio (2.5.15), it is possible to find:

$$N = -\frac{2}{M} \ln\left|\frac{E}{E_0}\right| = -\frac{2}{M} \ln\left|\frac{H}{H_0}\right|. \tag{2.5.26}$$

With the purpose of using a uniform average number of filling for excitation of atoms and turn of its magnetic moment in a magnetic field of impulse, we believe a two-level quantum system on magnetic quantum numbers also exists. In this case, size N gives a measure of the inversion of the system of atoms - radiators on the excited atoms [12], and also a measure of inversion of the system of the magnetic moments of atoms on magnetic quantum numbers.

We consider the dependence of the average number of filling on the time $N(t)$. If we are to accept the proportion of polarization of separate bi-level atoms to the electric field strength in the impulse, then, in accordance with the Maxwell-Bloch equations, the average atoms of the considered volume and the filling number is proportional to the volumetric density of the electromagnetic wave energy [13].

However, such a monotonous dependence between these values can not remain on the whole extent of the impulse.

Firstly, at the high volumetric density of impulse energy W, typical for SIT, when the impulse energy in the central part is higher than some certain size W, there exists the energetic saturation of the medium. The average filling number, thus, is $N = 1$, where all the atoms are excited, as shown in figure 2.5.2 (curve 1 - the dependence W on time, thicker curve 2 – used dependence N on time). The violation of proportion in the central part of the impulse is the basic drawback of this frequently-used system of the Maxwell-Bloch equations for the SIT description.

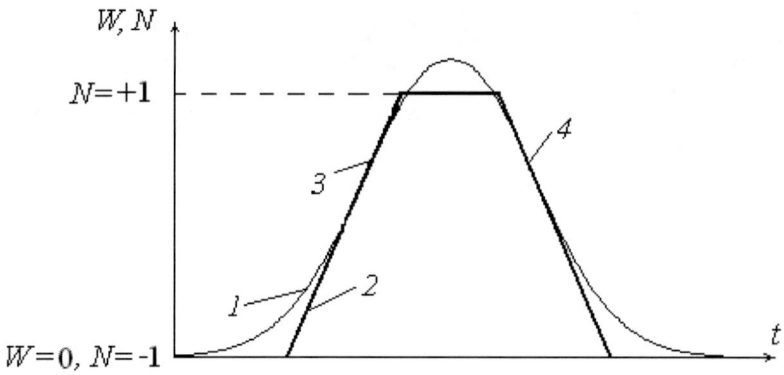

Figure 2.5.2.

Secondly, the period of value N relaxation is not less than [12], which is why the dependence cannot repeat high-frequently oscillations on both fronts of the impulse. The dependence could characterize only the proportion of average filling numbers and envelope W (curve 1) in the impulse. However, in two points of the fold (3 and 4 in figure 2.5.2) on the sites of increase and decrease of the envelope W , the value hence, $\partial^2 N / \partial t^2 = 0$ also. Besides, the dependence (a thick curve on figure 2.5.2) has the symmetrical character as at the SIT impulse becomes the conservative system (there is no reverse scattering or dissipation of energy). Therefore, it could be assumed with great accuracy that on the whole extent of impulse, except the points of the curve's fold, that the condition remains:

$$\frac{\partial^2 N}{\partial t^2} = 0, \tag{2.5.27}$$

while the dependence has the character as shown in figure 2.5.2, thick curve 2. The high generality of formula (2.5.27) could also be highlighted, which is possible for any piecewise linear function $N(t)$. Thus, the points of function break are excluded, as the derivates undergo the break.

The one-dimensional wave equation for the electric and magnetic components of the electromagnetic field for the considered problem is:

$$\frac{\partial^2}{\partial X^2}\begin{pmatrix} E \\ H \end{pmatrix} - \frac{1}{c^2}\begin{pmatrix} \mu \\ \varepsilon \end{pmatrix}\frac{\partial^2}{\partial t^2}\begin{pmatrix} E \\ H \end{pmatrix} = \frac{1}{c^2}\begin{pmatrix} \mu / \varepsilon_0 \\ \varepsilon \end{pmatrix}\frac{\partial^2}{\partial t^2}\begin{pmatrix} P \\ J \end{pmatrix}, \tag{2.5.28}$$

where or $E \equiv E_Z$, or $H \equiv H_Z$, J is the magnetization of the medium, - the speed of light in vacuum.

For electric component E the equation (2.5.28) is identical to (2.5.6). For magnetic component H the Maxwell equations (2.5.1) and (2.5.2) are as and $\mathrm{rot}\boldsymbol{E} = -\frac{\partial \boldsymbol{B}}{\partial t}$, and also a ratio for magnetization is used.

Similarly (2.5.10), we introduce the transformation of electric and magnetic field strength under the formula:

$$\begin{pmatrix} E(X,t) \\ H(X,t) \end{pmatrix} = \Phi(X,t)\exp(-i\omega_0 t), \tag{2.5.29}$$

where is a component of the cyclic frequency high-frequency oscillation of fields. Substituting (2.5.29) in (2.5.28), we obtain:

$$\left[\frac{\partial^2 \Phi}{\partial X^2} - \frac{1}{c^2}\begin{pmatrix} \mu \\ \varepsilon \end{pmatrix}\left(\frac{\partial^2 \Phi}{\partial t^2} - 2i\omega_0 \frac{\partial \Phi}{\partial t} - \omega_0^2 \Phi\right)\right]\exp(-i\omega_0 t) = \\ = \frac{1}{c^2}\begin{pmatrix} \mu/\varepsilon_0 \\ \varepsilon \end{pmatrix}\frac{\partial^2}{\partial t^2}\begin{pmatrix} P \\ J \end{pmatrix}. \tag{2.5.30}$$

We estimate the relative size of the first and second addends in the round brackets of the left side (2.5.30). For this purpose we introduce the scales of values time t and Φ:

$$T = t/t^*, \quad \Phi_0 = \Phi/\Phi^*,$$

where the asterisk designates dimensionless parameters. For the time scale, the duration (period) of impulse T should be logically chosen. The scale is chosen from the condition that the dimensionless second derivative and the dimensionless function are in the same order. Hence, the first addend in round brackets (2.5.30) is $\frac{\Phi_0}{T^2}\frac{\partial^2 \Phi^*}{\partial t^{*2}}$, and the last one $\omega_0^2 \Phi_0 \Phi^*$.

Instead of impulse period T we introduce a cyclic frequency of impulse $\omega = \frac{2\pi}{T}$. By comparing these addends, it is realized that as the cyclic frequency of impulse is of far less frequency than that of the field's oscillations, especially when $\omega^2 << \omega_0^2$. Similarly, it can be presented that the second addend in the round brackets (2.5.30) is far greater than the first one.

Hence, by disregarding the first small addend in the round brackets in the left part of (2.5.30), we find:

$$\left[\frac{\partial^2 \Phi}{\partial X^2} - \frac{1}{c^2}\begin{pmatrix}\mu\\ \varepsilon\end{pmatrix}\left(2i\omega_0 \frac{\partial \Phi}{\partial t} + \omega_0^2 \Phi\right)\right]\exp(-i\omega_0 t) = \\ = \frac{1}{c^2}\begin{pmatrix}\mu/\varepsilon_0\\ \varepsilon\end{pmatrix}\frac{\partial^2}{\partial t^2}\begin{pmatrix}P\\ J\end{pmatrix} . \qquad (2.5.31)$$

Assuming a vector of polarization P or magnetization J is directly proportional, accordingly, the electric and magnetic fields' strength, it is possible from (2.5.28) to receive at once the wave equation which the wave of any form satisfies. But there is the physical mechanism creating a restriction of the wave form. This mechanism is connected to the reradiated path of the atomic electron the electromagnetic impulse. We shall consider this process in more detail.

Let's write down the electric and magnetic fields' strength of the impulse as:

$$\begin{pmatrix}E(X,t)\\ H(X,t)\end{pmatrix} = \begin{pmatrix}|E(X,t)|\\ |H(X,t)|\end{pmatrix}\exp[i(rX - \delta t)], \qquad (2.5.32)$$

where r is the wave number and δ - bearing wave frequency of filling - constants, and - modules of functions and $H(X,t)$.

Formulas (2.5.26) and (2.5.27) reflect an offered physical model of interaction of electric and magnetic fields of the impulse with atoms at SIT.

Hence, taking into account (2.5.26), from (2.5.27) we have:

$$\frac{\partial^2 \ln\left|\frac{E}{E_0}\right|}{\partial t^2} = \frac{\partial^2 \ln\left|\frac{H}{H_0}\right|}{\partial t^2} = 0 . \qquad (2.5.33)$$

Transforming (2.5.33), we find:

$$\frac{\partial^2 |E|}{\partial t^2} = \left(\frac{\partial \ln|E|}{\partial t}\right)^2 |E| . \qquad (2.5.34)$$

A similar ratio takes place for function $|H|$. The given ratio cannot be accepted as the equation for determination of the module of electric and magnetic components of impulse. They are the approximated expression of the second derivative or for the examined physical model and reflects some nonlinear effects of the interaction between the electromagnetic radiation and the substance. The approximated ratio (2.5.34) determines the connection of medium polarization P with the electric field strength of impulse (similarly magnetizations J with the magnetic field strength), that will be shown further, see (2.5.36). It is necessary to calculate the electromagnetic field strength of impulse proceeding from the equation (2.5.28) taking into account the ratio (2.5.34).

According to (2.5.32) we have:

$$\begin{pmatrix} |E(X,t)| \\ |H(X,t)| \end{pmatrix} = \begin{pmatrix} E(X,t) \\ H(X,t) \end{pmatrix} \exp[-i(rX - \delta t)],$$

Hence, from (2.5.34) for an electric field of the impulse we have:

$$\frac{\partial^2 E}{\partial t^2} = -2i\delta \frac{\partial E}{\partial t} + \left[\delta^2 + \left(\frac{\partial \ln \left| \frac{E}{E_0} \right|}{\partial t} \right)^2 \right] E. \quad (2.5.35)$$

The ratio is similar for a magnetic field. Transiting in (2.5.35) to function under the formula (2.5.29) and using a condition where χ is a relative dielectric susceptibility of substance, we obtain:

$$\frac{\partial^2 P}{\partial t^2} = \left[\begin{matrix} -2i\delta\varepsilon_0 \chi \frac{\partial \Phi}{\partial t} - 2\delta\varepsilon_0 \chi \omega_0 \Phi + \\ + \varepsilon_0 \chi \left[\delta^2 + \left(\frac{\partial \ln \left| \frac{\Phi}{|\Phi_0|} \right|}{\partial t} \right)^2 \right] \Phi \end{matrix} \right] \exp(-i\omega_0 t), \quad (2.5.36)$$

For value $\frac{\partial^2 J}{\partial t^2}$, using a condition where χ is a relative magnetic susceptibility of substance, we obtain the similar ratio (2.5.36), with the exception that in the right part is absent.

Values $\begin{pmatrix} E_0 \\ H_0 \end{pmatrix} = \Phi_0 \exp(-i\omega_0 t)$. Comparing (2.5.29) and (2.5.32), we have $\begin{pmatrix} |E| \\ |H| \end{pmatrix} = |\Phi|$, $\begin{pmatrix} |E_0| \\ |H_0| \end{pmatrix} = |\Phi_0| = const$.

Substituting (2.5.36) in (2.5.31) we find:

$$2i(\omega_0+\chi\delta)\frac{\partial\Phi}{\partial t}+c^2\begin{pmatrix}1/\mu\\1/\varepsilon\end{pmatrix}\frac{\partial^2\Phi}{\partial X^2}+\left(\omega_0^2+2\chi\delta\omega_0-\chi\delta^2\right)\Phi=$$

$$=\chi\left(\frac{\partial\ln\left|\frac{\Phi}{\Phi_0}\right|}{\partial t}\right)^2\Phi \qquad (2.5.37)$$

In the equation (2.5.37) the value χ is a relative dielectric susceptibility for an electric field and a relative magnetic susceptibility for magnetic components of electromagnetic field.

The nonlinear Schrödinger equation with a rather complicated form of nonlinearity is obtained. We shall enter the following designations:

$$\alpha=\omega_0+\chi\delta,\quad \gamma=\omega_0^2+2\chi\delta\omega_0-\chi\delta^2=\alpha^2-\begin{pmatrix}\varepsilon\\\mu\end{pmatrix}\chi\delta^2,$$

where is the relative permeability of the substance. Hence, the equation (2.5.37) will written as:

$$2i\alpha\frac{\partial\Phi}{\partial t}+c^2\begin{pmatrix}1/\mu\\1/\varepsilon\end{pmatrix}\frac{\partial^2\Phi}{\partial X^2}+\gamma\Phi=\chi\left(\frac{\partial\ln\left|\frac{\Phi}{\Phi_0}\right|}{\partial t}\right)^2\Phi. \qquad (2.5.38)$$

Let's search the solution of the nonlinear Schrödinger equation (2.5.38) as:

$$\Phi=\Phi_0 f(kX-\omega t)\exp\left[i\left(rX-\delta^* t\right)\right], \qquad (2.5.39)$$

where the form of the function is yet unknown. The wave number of impulse k, and also frequencies of impulse ω and waves of filling , are constants. Having designated $\zeta=kX-\omega t$ and having substituted (2.5.39) in (2.5.38) and also taking into account $|\Phi|=\Phi_0 f(\zeta)$, we obtain:

$$c^2\begin{pmatrix}1/\mu\\1/\varepsilon\end{pmatrix}k^2\frac{d^2 f}{d\zeta^2}+2i\frac{df}{d\zeta}\left(krc^2\begin{pmatrix}1/\mu\\1/\varepsilon\end{pmatrix}-\alpha\omega\right)+$$

$$+f\left(\gamma+2\alpha\delta^*-r^2c^2\begin{pmatrix}1/\mu\\1/\varepsilon\end{pmatrix}\right)==\chi\omega^2\left(\frac{d\ln f}{d\zeta}\right)^2 f \qquad (2.5.40)$$

If we are to accept $krc^2 = \alpha\begin{pmatrix}\mu\\ \varepsilon\end{pmatrix}\omega$, since in (2.5.40) there should not be imaginary addends, this equation will be transformed to the form:

$$c^2\begin{pmatrix}1/\mu\\ 1/\varepsilon\end{pmatrix}k^2\frac{d^2 f}{d\zeta^2}+f\left(\gamma+2\alpha\delta^* - r^2c^2\begin{pmatrix}1/\mu\\ 1/\varepsilon\end{pmatrix}\right)=\chi\omega^2\left(\frac{d\ln f}{d\zeta}\right)^2 f \cdot \tag{2.5.41}$$

Let's search the solution to the equation (2.5.41) as:

$$f = C_1\exp\left[\frac{C_2\zeta^2}{4}\right], \tag{2.5.42}$$

where and are constants. Having substituted (2.5.42) in (2.5.41), we find that the constant can be any size and $\chi\begin{pmatrix}\mu\\ \varepsilon\end{pmatrix}\omega^2 = k^2c^2$.

The constant cannot depend on parameters of the equation. We accept that $C_2 = -1$. Then the frequency and wave number in (2.5.39), accordingly, are equal:

$$\delta^* = \frac{c^2\begin{pmatrix}1/\mu\\ 1/\varepsilon\end{pmatrix}\left(r^2+k^2/2\right)-\gamma}{2\alpha};\ r = \frac{\alpha\begin{pmatrix}\mu\\ \varepsilon\end{pmatrix}\omega}{kc^2}. \tag{2.5.43}$$

The formulas in (2.5.43) connect the frequency and wave number oscillations of functions with the parameters of substance and electromagnetic field of impulse.

The most simple ratio between parameters arises if we put $\delta = \omega_0$.

In this case and $\gamma = \begin{pmatrix}\varepsilon\\ \mu\end{pmatrix}\delta^2$. From the ratio (2.5.43), taking into account $\chi\begin{pmatrix}\mu\\ \varepsilon\end{pmatrix}\omega^2 = k^2c^2$, we find:

$$r = \frac{\alpha\begin{pmatrix}\mu\\ \varepsilon\end{pmatrix}\omega}{kc^2} = \frac{\alpha k}{\chi\omega} = \frac{\begin{pmatrix}\varepsilon\\ \mu\end{pmatrix}\delta k}{\chi\omega},$$

$$\delta^* = \frac{\alpha}{2\chi} + \frac{\chi\omega^2}{4\alpha} - \frac{\gamma}{2\alpha} = \frac{2\begin{pmatrix}\varepsilon\\ \mu\end{pmatrix}\delta^2 + \chi^2\omega^2}{4\chi\begin{pmatrix}\varepsilon\\ \mu\end{pmatrix}\delta}. \tag{2.5.44}$$

Taking into account, that $2\begin{pmatrix}\varepsilon\\ \mu\end{pmatrix}\delta^2 >> \chi^2\omega^2$, we have $\delta^* \approx \frac{\delta}{2\chi}$. The used inequality is correct, since for rarefied gas ($n_0 < 10^{18}\ atoms/cm^3$) the values and wave frequency of filling of impulse δ is much greater than the frequencies of the impulse envelope ω.

Taking into account (2.5.32), (2.5.42) and the condition $\begin{pmatrix}|E|\\ |H|\end{pmatrix} = |\Phi|$, it is possible to write down the laws of change of electromagnetic field strength as:

$$\begin{pmatrix}E\\ H\end{pmatrix} = \begin{pmatrix}E_0\\ H_0\end{pmatrix}\exp\left[-\frac{(kX-\omega t)^2}{4}\right]\exp[i(rX-\delta t)]. \tag{2.5.45}$$

It is interesting to note that although the ratios for an electric component of impulse in (2.5.24) and (2.5.45) are similar and have identical phases of oscillations which are correct some distance from the reradiated atom, Schrödinger's nonlinear equations differ in their type of nonlinearity. It is connected in paragraph 2.5.1 where the impulse concerning the low intensity was considered, not resulting in the energy saturation of the medium in which it is propagated.

For calculation purposes, similarly to paragraph 2.5.1, we use

$$k = \frac{\omega}{V} = 2.1\cdot 10^4 m^{-1},\ r = \frac{\delta}{c} = 2.1\cdot 10^5 m^{-1},$$
$$\omega = 6.28\cdot 10^{12}\ s^{-1},\ \delta = 6.28\cdot 10^{13}\ s^{-1}.$$

The result of the calculation, for example, of the electric field strength of impulse on coordinate X, corresponds in accuracy to figure 2.5.1.

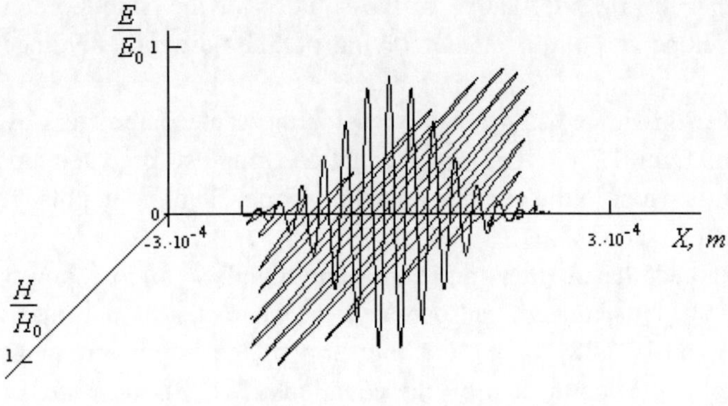

Figure 2.5.3.

Taking into account the perpendicularity of planes of oscillations of vectors of electric and magnetic fields in the impulse, it is possible to obtain a form of electromagnetic soliton, as shown in figure 2.5.3.

In figure 2.5.4, for comparison, we see the envelope of an impulse of an electric field in case SIT, plotted under the formula (2.5.45), a curve 1, and under the formula (2.5.46) being an investigation of Maxwell-Bloch's theory, curve 2. The envelope of an impulse of electric field strength in this theory is formed as the first derivative from solving the Sine-Gordon equation and appears thus:

$$E = \frac{E_0}{\mathrm{ch}(kX - \omega t)}. \qquad (2.5.46)$$

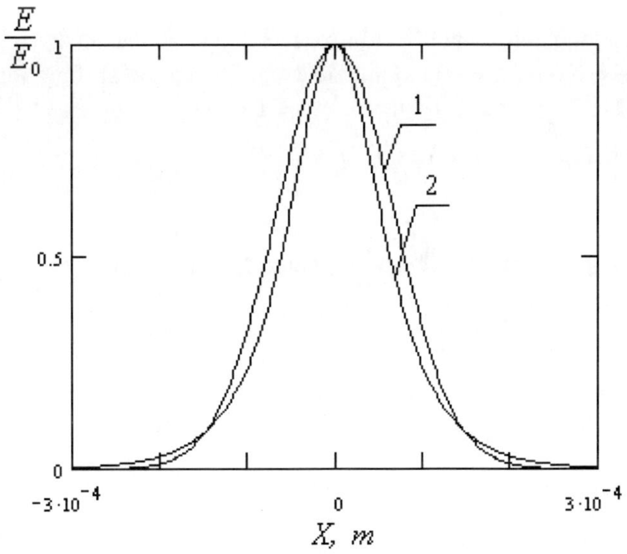

Figure 2.5.4.

It is known that the first derivative of the solution of the Sine-Gordon equation looks similar to the envelope of soliton in the solving of Schrödinger's nonlinear equation with cubic nonlinearity.

Curves 1 and 2 on figure 2.5.4 are plotted for the same parameters as the function in figure 2.5.1. From figure 2.5.4 it is visible that the impulse received under the formula (2.5.45), a curve 1, is wider in the average area, but shorter than the impulse received with the help of Maxwell-Bloch, curve 2.

It should be noted that at the transition from impulses of low power to high-power impulses including a self-induced transparency, the character of nonlinearity of the basic equations (2.5.20) and (2.5.38) varies, but the form of the impulse remains identical - see (2.5.24) and (2.5.45). It is connected by the equations (2.5.20) and (2.5.38) which have an identical one-soliton solution.

2.5.3. Some Features of Schrödinger's Nonlinear Equations Solutions

It is known [14] that the solution of Schrödinger's nonlinear equation with cubic nonlinearity represents the soliton of envelope as a return hyperbolic cosine. To find out what restrictions should satisfy the velocity characteristics as waves of filling of an impulse and impulse envelope, we shall consider in more detail the finding of one-soliton solving of this equation:

$$i\frac{\partial \varphi}{\partial t}+\frac{\lambda}{2}\frac{\partial^2 \varphi}{\partial X^2}=-\omega|\varphi|^2\varphi, \tag{2.5.47}$$

where is a function of the coordinate X and time t, λ and ω - constants.

Let's search the solving (2.5.47) having presented the unknown function as:

$$\varphi(X,t)=F(X,t)\exp i(rX-\delta t), \tag{2.5.48}$$

where is the module of function $\varphi(X,t)$, r and δ - accordingly wave number and cyclic frequency of an impulse filling wave.

Having substituted (2.5.48) in (2.5.47), we shall obtain one-soliton solution of the equation (2.5.47) as:

$$\varphi(X,t)=\sqrt{\frac{k}{r}}\frac{1}{\mathrm{ch}(kX-\omega t)}\exp i(rX-\delta t)\cdot \tag{2.5.49}$$

where k and ω are accordingly wave number and cyclic frequency of envelope of an impulse.

Thus, conditions should be observed:

$$\omega = \lambda rk \text{ and } \delta = \frac{\lambda}{2}\left(r^2 - k^2\right). \tag{2.5.50}$$

Let's find the ratio of frequencies of a filling wave and impulse envelope:

$$\frac{\delta}{\omega} = \frac{1}{2}\left(\frac{r^2 - k^2}{rk}\right) = \frac{1}{2}\left(\frac{r}{k} - \frac{k}{r}\right). \tag{2.5.51}$$

Deciding the equation (2.5.51) to be relative $\frac{r}{k}$, we obtain:

$$\frac{r}{k} = \frac{\delta}{\omega} \pm \sqrt{\left(\frac{\delta}{\omega}\right)^2 + 1}. \tag{2.5.52}$$

In the formula (2.5.52), physical senses without the minus sign, since . From (2.5.52), it follows that $\frac{r}{k} > 2\frac{\delta}{\omega}$. This condition can be transformed, taking into account that is a velocity of the impulse envelope, and is a velocity of a filling wave. Thus,

$$V_{env} > 2V_f. \tag{2.5.53}$$

The condition (2.5.53) results in the restriction of an area of the physical problems described by the equation (2.5.47). The formula (2.5.53) shows that the equation (2.5.47) cannot be used frequently, for example, for the modelling of propagation of electromagnetic impulse in dielectric. Really, for this phenomenon a small velocity of the impulse can be a characteristic at high velocity of electromagnetic filling wave.

There is a question as to whether there are nonlinear equations similar (2.5.47), for which individual solution is not limited to an inequality (2.5.53).

Let's consider the following Schrödinger's nonlinear equation with positive cubic nonlinearity:

$$i\frac{\partial\varphi}{\partial t} + \frac{\lambda}{2}\frac{\partial^2\varphi}{\partial X^2} = \omega|\varphi|^2\varphi.$$

This equation differs from (2.5.47) only by the sign to the right part. As well as in the previous case, we shall search for the solution as (2.5.48). This solution looks like this:

$$\varphi(X,t) = \sqrt{\frac{k}{r}}\frac{1}{\cos(kX - \omega t)}\exp i(rX - \delta t).$$

The condition should also satisfy:

$$\omega = \lambda rk \text{ and } \delta = \frac{\lambda}{2}\left(r^2 + k^2\right).$$

Obviously, the received solution has no form of a solitary wave.

Analyzing this, as well as in the previous case, we find a ratio between the velocities offilling wave and impulse envelope:

$$V_{env} < 2V_f .$$

In spite of the fact that the received inequality is opposite to an inequality (2.5.53), the considered equation, naturally, cannot be used for the solving physical problems in which there are solitary waves.

Let's consider the following nonlinear differential equation:

$$i\frac{\partial\varphi}{\partial t} + \frac{\lambda}{2}\frac{\partial^2\varphi}{\partial X^2} = -\mu \ln\left|\frac{\varphi}{\varphi_0}\right|\varphi, \tag{2.5.54}$$

where μ is a constant.

We shall search for the solution of this equation as (2.5.48), using an argument $\zeta = kX - \omega t$. For the function , we shall derive the differential equation:

$$\frac{\lambda k^2}{2}F_{\zeta\zeta} + \left(\delta - \frac{\lambda r^2}{2}\right)F = -\mu\left(\ln\frac{F}{F_0}\right)F \cdot \tag{2.5.55}$$

In equation (2.5.55), the value of $F_0 = \varphi_0$. As before, the condition should be satisfied. To solve equation (2.5.55) we search the function of Gauss:

$$F = F_0 \exp\left(-C\zeta^2\right), \tag{2.5.56}$$

where C is the unknown constant of integration.

Substituting (2.5.56) in (2.5.55), we obtain:

$$\frac{\lambda k^2}{2}2C\left(2C\zeta^2 - 1\right) + \left(\delta - \frac{\lambda r^2}{2}\right) = \mu C\zeta^2 \cdot \tag{2.5.57}$$

Equating in (2.5.57) addends at identical degrees ζ^2 , we find:

$$C = \frac{\mu}{2\lambda k^2} \text{ and } \delta = \frac{\lambda}{2}\left(r^2 + \frac{\mu}{\lambda}\right). \qquad (2.5.58)$$

If we are to assume that $C = 1/2$, whence it follows $\mu = \lambda k^2$, we obtain the formula $\delta = \frac{\lambda}{2}\left(r^2 + k^2\right)$. Hence, the ratio of velocities of impulse envelope and velocity filling wave for the equation (2.5.54) satisfies the inequality:

$$V_{env} < 2V_f. \qquad (2.5.59)$$

The inequality (2.5.59) shows that the equation (2.5.54) can be used for the description of the phenomenon of occurrence of electric impulse in the substance that has been used in paragraph 2.5.1.

The one-soliton solution of the equation (2.5.54) according to (2.5.48) and (2.5.56) is function:

$$\varphi(X,t) = F_0 \exp\left[-\frac{(kX-\omega t)^2}{2}\right]\exp i(rX - \delta t). \qquad (2.5.60)$$

Let's consider the nonlinear Schrödinger equation in which nonlinearity has a more non-stationary character than in the equations (2.5.47) and (2.5.54):

$$i\frac{\partial\varphi}{\partial t} + \frac{\lambda}{2}\frac{\partial^2\varphi}{\partial X^2} = \eta\left(\frac{\partial \ln|\varphi|}{\partial t}\right)^2 \varphi, \qquad (2.5.61)$$

where η is a constant.

To solve this equation we shall search (2.5.48), using an argument $\zeta = kX - \omega t$. For the function we shall obtain the differential equation:

$$\frac{\lambda k^2}{2}F_{\zeta\zeta} + \left(\delta - \frac{\lambda r^2}{2}\right)F = \eta\omega^2\left[(\ln F)_\zeta\right]^2 F. \qquad (2.5.62)$$

As well as in the previous cases, the condition should be satisfied. The solving of the equation (2.5.62) we search as a function of Gauss (2.5.56). Substituting (2.5.56) in (2.5.62) we obtain:

$$\frac{\lambda k^2}{2} 2C\left(2C\zeta^2 - 1\right) + \left(\delta - \frac{\lambda r^2}{2}\right) = 4\eta\omega^2 C^2 \zeta^2. \qquad (2.5.63)$$

Equating in (2.5.63) addends at identical degrees ζ^2, we find:

$$\eta = \frac{\lambda k^2}{2\omega^2} \text{ and } \delta = \frac{\lambda}{2}\left(r^2 + 2Ck^2\right). \qquad (2.5.64)$$

If we are to assume that , we deduce the formula $\delta = \frac{\lambda}{2}\left(r^2 + k^2\right)$. The ratio between velocities of waves of filling and impulse envelope corresponds to (2.5.59). Hence, the equation (2.5.61) can describe the physical processes of propagation of electromagnetic impulses in the substance that has been used in paragraph 2.5.2. The one-soliton solving of the equation (2.5.61) is completely similar (2.5.60).

Thus, the equations (2.5.54) and (2.5.61) can be essentially used for the description of the modulated electromagnetic impulse in substance, as against the equation with cubic nonlinearity (2.5.47). This circumstance was taken into account for finding the form of electromagnetic impulse in paragraphs 2.5.1 and 2.5.2.

Chapter 3

THE EFFECTIVE SECTION OF THE INTERACTION OF AN ELECTROMAGNETIC FIELD WITH A SUBSTANCE

3.1. EFFECTIVE SECTION OF INTERACTION OF ELECTROMAGNETIC RADIATION WITH A PARTICLE

The concept of effective section of interaction (absorption or scattering) electromagnetic radiation with a particle of substance is widely used in the physics of interaction processes.

This value characterizes the probability of interaction. The effective section is calculated for the diversified phenomena: absorption of radiation, elastic and non-elastic scattering, ionization, recharge of particles, suppression of a luminescence, rotation of a polarization plane of light by a particle, etc. The effective section of interaction extremely powerful method of analysis of interaction of electromagnetic radiation with substance there are especially found by using the quantum approach. But also with the classical approach, which will be considered in this chapter, the given method of analysis yields good, informative results.

***

Numerically, the effective section of interaction is equal to the area with a center that coincides with the center of a particle. However, depending on the probability of interaction, the effective section σ can be greater or less or equal to the geometrical area of a section of a particle. Dimension $[\sigma]=m^2$. Let's consider volume V, figure 3.1.1, in which there are particles with numerical concentration:

$$n_0 = \frac{\Delta N}{\Delta \mathrm{V}}, \qquad (3.1.1)$$

where is the number of particles in volume $\Delta \mathrm{V}$.

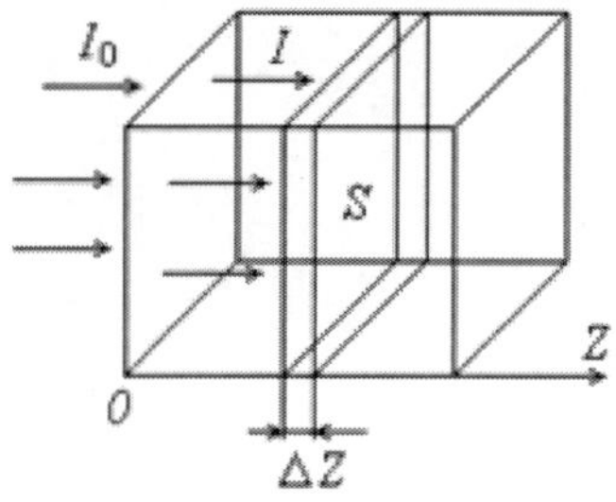

Figure 3.1.1.

Total effective section of interaction of all particles in volume equally:

$$\Delta S = \sigma \Delta N = \sigma n_0 \Delta V = \sigma n_0 S \Delta Z \, . \tag{3.1.2}$$

In writing the last formula, we have assumed that particles do not overlap each other. It is equivalent to a small concentration of particles. If the concentration of particles is great due to overlapping, the portion of particles will not take part in the absorption of light.

Let's estimate from the classical point of view the connection of the totally effective section of interaction with the geometrical area of cross-section of all absorbing particles provided that at the most dense packing particles all electromagnetic radiation is absorbed. We shall consider what minimal area the molecules of substance in the cross-section can occupy. We shall assume that molecules have an approximately spherical form.

In figure 3.1.2 the dense packing of molecules of substance in diameter *d,* which completely absorbs all radiation, is shown. Examining two consecutive molecules, we notice that there is an area which is inaccessible to the center of the next molecule. This inaccessible area equals $S_{in} = \pi d^2 = 4S_1$, where - the geometrical area of cross-section of one molecule. If the examined unit of volume of substance is molecules, the full inaccessible area equals $\Delta S = 4n_0 S_1$. Thus, the total effective section of absorption is equal to the area of cross-section of all molecules. Below, this result will be presented in a more formal way.

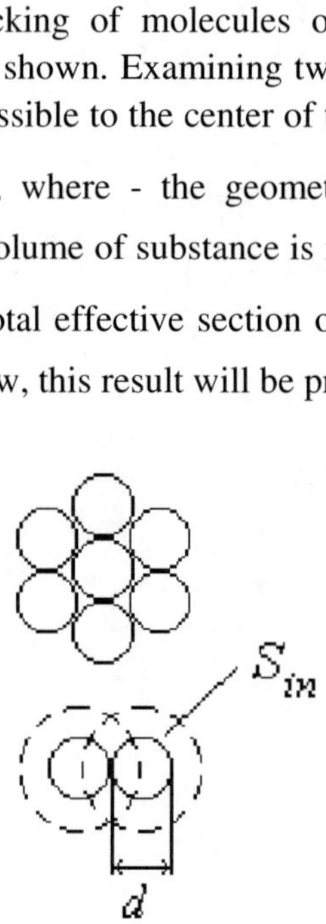

Figure 3.1.2.

As the portion of the effective section of interaction for all particles in the general section *S* is proportional to a portion absorbed or scattered on thickness of intensity of light *I* it is possible to write down:

$$\frac{\Delta S}{S} = -\frac{\Delta I}{I} = \sigma n_0 \Delta Z \quad , \qquad (3.1.3)$$

where the minus sign specifies a reduction of intensity of light with an increase in coordinate *Z*. With the growth of total effective section of interaction, $\Delta S > 0$, the intensity of light passing through decreases $\Delta I < 0$. The formula (3.1.3) is Bouguer's law in the differential form. Integrating, we find $I = I_0 \exp(-\alpha Z)$, where a parameter of absorption $\alpha = \sigma n_0$, see (2.5.14). The much greater interest represents the distribution of intensity of radiation on angles after its interaction with atoms (first of all scattering).

In this case the differential effective section of scattering is entered:

$$d\sigma = \frac{dP}{I_0}. \qquad (3.1.4)$$

The differential effective section of scattering is numerically equal to the ratio of a flux of energy of radiation dP, a scattered particle in a solid angle $d\Omega$, to the intensity of the radiation falling on a particle, figure 3.1.3.

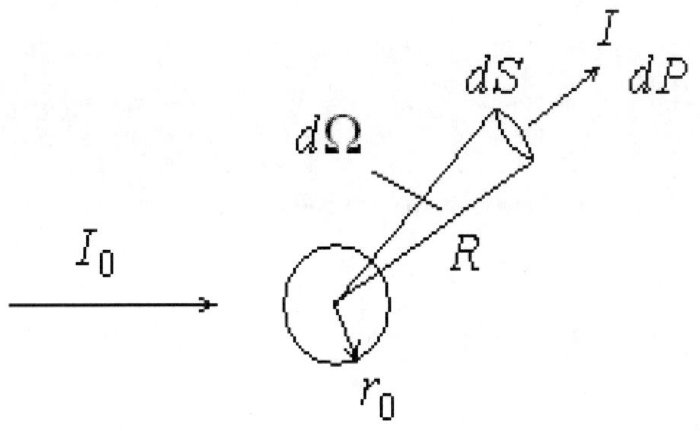

Figure 3.1.3.

Taking into account the connection of intensity with a flux of energy $I = \dfrac{dP}{dS}$, where is the area through which the given flux of energy passes, the formula (3.1.4) can be copied as:

$$d\sigma = \frac{IdS}{I_0} = \frac{IR^2 d\Omega}{I_0}, \tag{3.1.5}$$

where a determination of a solid angle is used, *R* is the radius of any sphere surrounding a particle on which some area is conditionally cut out.

If the particle has the spherical form (even conditionally), that, reducing the radius $R \to r_0$, it is possible to use the radius of the particle r_0. Since a particle which scatters is more often electron, the classical radius of electron is used. It is defined from the equality of the energy of the electron rest and the energy of the electron electric field on its surface, arising due to its charge *e*.

In this case the formula (3.1.5) allows us to find a connection between the intensity scattered on electron radiations and the intensity of radiation falling on a particle:

$$I = I_0 \frac{d\sigma}{r_0^2 d\Omega}. \tag{3.1.6}$$

If, by analogy to paragraph 2.5.2, we accept the probability of scattering of radiation , the probable essence of effective section of scattering becomes clear. At the equal probability of scattering in all directions $p = const$, taking into account the geometrical ratio, we have $\sigma = Sp$, where is the area of a surface of a scatter particle - electron or, more physically, the area of its cross-section.

It should also be noted that the parameter of absorption of the substance is connected to the probability of absorption of radiation by a particle under the formula $\alpha = n_0\sigma = n_0 Sp$, i.e. the parameter of absorption of the substance is equal to the total area of the cross-section of all absorbing particles in the unit of volume of the substance multiplied by the probability of absorption of radiation by one particle.

3.1.1. Dipole Radiation

Before proceeding to a further analysis of the effective sections of interaction of electromagnetic radiation with various electronic systems, we shall briefly consider the concept of dipole radiations. Oscillating electric dipole *D* radiates an electromagnetic wave. Let's consider the radiation of this wave in a direction of an unit vector $\boldsymbol{n} = \frac{\boldsymbol{r}}{r}$, figure 3.1.4. Vectors of electric field strength ***E***, magnetic field ***H*** and a radius-vector ***r*** in a point of observation form a right-spiral system.

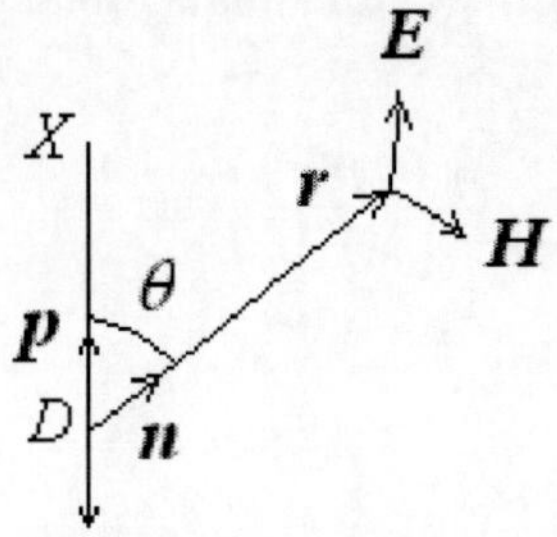

Figure 3.1.4.

The solving of Maxwell's equations in vacuum:

$$\mathrm{rot}\boldsymbol{H} = \varepsilon_0 \frac{\partial \boldsymbol{E}}{\partial t}, \tag{3.1.7}$$

$$\mathrm{rot}\boldsymbol{E} = -\mu_0 \frac{\partial \boldsymbol{H}}{\partial t}, \tag{3.1.8}$$

$$\mathrm{div}\boldsymbol{E} = 0, \tag{3.1.9}$$

$$\mathrm{div}\boldsymbol{H} = 0, \tag{3.1.10}$$

for an oscillating dipole is well-known, for example [21].

This solution looks like:

$$\boldsymbol{H} = \frac{1}{4\pi cr}\left[\frac{\partial^2 \boldsymbol{p}(\tau)}{\partial t^2} \times \boldsymbol{n}\right], \tag{3.1.11}$$

$$\boldsymbol{E} = \mu_0 c[\boldsymbol{H} \times \boldsymbol{n}], \tag{3.1.12}$$

where is a dipole moment of an oscillating dipole *D*. Argument $\tau = t - \frac{r}{c}$, where is the time of delay of the oscillations of a field in a point of observation in comparison with the oscillations of a dipole.

Passing to modules of electric and magnetic fields' strength, we have:

$$H = \frac{1}{4\pi cr}\frac{\partial^2 p(\tau)}{\partial t^2}\sin\theta, \tag{3.1.13}$$

$$E = \mu_0 cH = \frac{\mu_0}{4\pi r}\frac{\partial^2 p(\tau)}{\partial t^2}\sin\theta, \tag{3.1.14}$$

where θ is an angle between vectors $\boldsymbol{p}$ and $\boldsymbol{r}$.

The volumetric density of energy of radiation is found in the formula:

$$W = \frac{\varepsilon_0 E^2}{2} + \frac{\mu_0 H^2}{2} = \frac{\mu_0}{16\pi^2 c^2 r^2}\left(\frac{\partial^2 p(\tau)}{\partial t^2}\right)^2 \sin^2\theta, \tag{3.1.15}$$

and intensity of radiation:

$$I = Wc = \frac{\mu_0}{16\pi^2 c r^2}\left(\frac{\partial^2 p(\tau)}{\partial t^2}\right)^2 \sin^2\theta. \tag{3.1.16}$$

Furthermore, we need a flux of energy through some area dS, see figure 3.1.3:

$$dP = IdS = \frac{\mu_0}{16\pi^2 c r^2}\left(\frac{\partial^2 p(\tau)}{\partial t^2}\right)^2 \sin^2\theta\, r^2 d\Omega = \tag{3.1.17}$$

$$= \frac{\mu_0}{16\pi^2 c}\left(\frac{\partial^2 p(\tau)}{\partial t^2}\right)^2 \sin^2\theta d\Omega$$

If we are to assume that the oscillating dipole is an electronic shell of atom using (1.2.2), we shall find:

$$\frac{\partial^2 \boldsymbol{p}}{\partial t^2} = -\frac{\frac{e^2}{m}\boldsymbol{E}_0\omega^2 e^{i\omega t}}{\omega_0^2 - \omega^2 + i\gamma\omega}. \tag{3.1.18}$$

Hence:

$$dP = \frac{\mu_0}{16\pi^2 c}\left(\frac{\frac{e^2}{m}E_0\omega^2 e^{i\omega t}}{\omega_0^2 - \omega^2 + i\gamma\omega}\right)^2 \sin^2\theta d\Omega = , \tag{3.1.19}$$

$$= \frac{r_0^2}{\mu_0 c}\left(\frac{E\omega^2}{\omega_0^2 - \omega^2 + i\gamma\omega}\right)^2 \sin^2\theta d\Omega$$

where is the classical radius of electron and $E = E_0 e^{i\omega t}$.

3.2. Effective Section of Interaction of Electromagnetic Radiation with Free Electron

If an electromagnetic wave falls on a free electron the result compels the oscillations of the electron under action of an electric field of this wave, i.e. arise dipole - oscillator.

We shall consider the radiation of oscillating dipole *D*, figure 3.2.1a, arising under action by electromagnetic radiation intensity I_0. We shall neglect the action of a magnetic field of a falling wave.

Let's firstly assume that dipole *D* does not have its own frequency of oscillations $\omega_0 = 0$. There is also no attenuation of oscillations $\gamma = 0$. Therefore, such a dipole cannot be assumed to be an adequate model of the atom.

According to (3.1.19), we shall find a flux of energy of scattered radiation in a solid angle $d\Omega$, figure 3.2.1b:

$$dP = \frac{r_0^2}{\mu_0 c} E^2 \sin^2 \theta d\Omega = r_0^2 \varepsilon_0 c E^2 \sin^2 \theta d\Omega = \\ = r_0^2 c \frac{\varepsilon_0 E_0^2}{2} \sin^2 \theta d\Omega = r_0^2 I_0 \sin^2 \theta d\Omega \tag{3.2.1}$$

As the volumetric density of energy of electric and magnetic components in a wave in vacuum are equal $\frac{\varepsilon_0 E^2}{2} = \frac{\mu_0 H^2}{2}$, see (3.1.15), the intensity $I = cW = c\varepsilon_0 E^2$. We can use an average for the period of intensity of falling radiation $I_0 = c\frac{\varepsilon_0 E_0^2}{2}$. The same average intensity reradiates in the direction of axis *Z* i.e. $I_Z = I_0$.

Let's assume also that the falling electromagnetic wave is polarized along an axis of a dipole, i.e. in plane *XZ*.

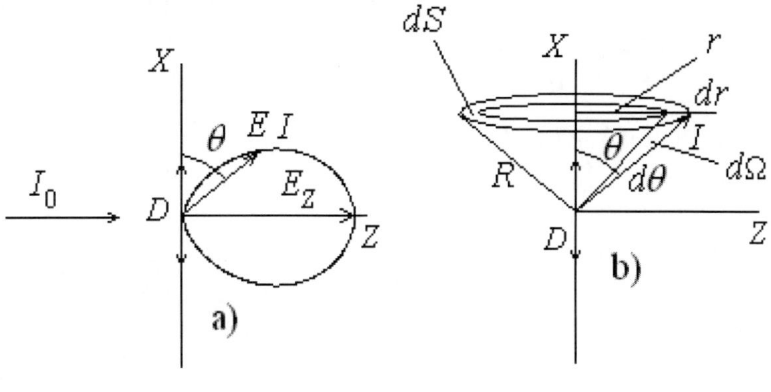

Figure 3.2.1.

Along an axis of dipole X reradiation (scattering) is absent, see (3.1.16). In the direction of axis Z it is maximum reradiation. The intensity of an electric field radiates under angle θ to an axis of a dipole equally $E = E_Z \sin\theta$, where is the intensity of the electric field, reradiating in the direction of axis Z.

In figure 1.3.2a the distribution of intensity $I = c\varepsilon_0 E^2 = c\varepsilon_0 E_Z^2 \sin^2\theta$, scattered electromagnetic waves are graphically shown depending on angle θ (actually, the vector is directed perpendicularly to axes Z). Dipole radiation is observed as well in the direction which is not shown in the figure.

Using (3.1.4) and (3.2.1), we shall find the differential effective section of the scattering of the polarized electromagnetic radiation on the free electron (on dipole D):

$$d\sigma = \frac{dP}{I_0} = r_0^2 \sin^2\theta d\Omega. \tag{3.2.2}$$

The intensity of scattered electromagnetic radiation, according to (3.1.6), equals:

$$I = I_0 \sin^2\theta. \tag{3.2.3}$$

Thus, in a direction of radiation falling on a dipole at , the initial intensity is kept. We can note also that the intensity of scattered radiation decreases in inverse proportion to a square of distance from a dipole $I \sim \frac{1}{r^2}$, see (3.1.16).

The formula (3.2.2) gives a differential effective section of light scattering polarized in plane XZ, on an oscillating electron, figure 3.2.1b, in a solid angle $d\Omega$

Let's find the full (total) effective section of scattering on dipole - oscillator D of the polarized electromagnetic radiation. In this case, it is necessary to integrate (3.2.2) into all of the space. By determining the solid angle where designations are given on figure 3.2.1b. The area is $dS = 2\pi r dr = 2\pi R \sin\theta R d\theta$. Hence, $d\Omega = 2\pi \sin\theta d\theta$. Therefore, the differential effective section of the scattering of radiation (3.2.2) can be written down as:

$$d\sigma = 2\pi r_0^2 \sin^3\theta d\theta. \tag{3.2.4}$$

Integrating into all of the space, we have:

$$\sigma = 2\pi r_0^2 \int_0^\pi \sin^3\theta d\theta = 2\pi r_0^2 \left(\frac{\cos^3\theta}{3} - \cos\theta \right)_0^\pi = \frac{8\pi}{3} r_0^2. \tag{3.2.5}$$

It should be noted that if the energy of falling radiation is great, for example, in an X-ray range, it is possible to neglect the coupling of electrons in atom and assume that the electrons of the atom scatter radiation as free particles [21].

Therefore, the received formula (3.2.5) frequently refers to Thomson's formula for the effective section of scattering on electrons. The numerical size is $\sigma = 6.65 \cdot 10^{-25}\, cm^2$. The experiment shows that Thomson's formula is correct in the substance for the lengths of waves $\lambda > 0{,}2\, nm$.

The formula (3.2.2) describes a two-dimensional case of light scattering polarized in a plane of radiation on the oscillating electron.

Let's find the differential effective section of scattering of non-polarized light by oscillating dipole *D*. Let's consider this dipole in space, figure 3.2.2. The ratio between the angles can be found, projecting on axes of coordinates a vector of intensity ***I***:

$$\cos\theta = \sin\delta\cos\varphi . \tag{3.2.6}$$

The rotation of coordinate axes on figure 3.2.2 is carried out for the presentation of geometry.

Using:

$$\sin^2\theta = 1 - \sin^2\delta\cos^2\varphi , \tag{3.2.7}$$

we shall find the average from the formula (3.2.7) on the azimuthal angle φ:

$$\sin^2\theta = 1 - \sin^2\delta < \cos^2\varphi > = 1 - \frac{\sin^2\delta}{2} = \frac{1+\cos^2\delta}{2} . \tag{3.2.8}$$

Averaging an azimuthal angle φ allows us to use the result (3.2.2) received for polarized light.

Hence, the differential effective section of the scattering for non-polarized light, according to (3.2.2), in a solid angle in the direction of axis *Z* under angle δ can be written as:

$$d\sigma = r_0^2 \sin^2\theta d\Omega = 2\pi r_0^2 \frac{1+\cos^2\delta}{2} \sin\delta d\delta , \tag{3.2.9}$$

where in this case the solid angle $d\Omega = 2\pi\sin\delta d\delta$, since scattering in a solid angle in a direction of axis *Z* is examined.

For the full section of scattering:

$$\sigma = 2\pi r_0^2 \int_0^{\pi} \frac{1+\cos^2\delta}{2} \sin\delta d\delta = \frac{8\pi}{3} r_0^2 . \tag{3.2.10}$$

As one would expect, the result is identical (3.2.5), i.e. the Thomson's formula.

For the reception of intensity of scattered non-polarized light, using and the formula (3.1.6), we shall find:

$$I = I_0 \frac{1+\cos^2\delta}{2}. \tag{3.2.11}$$

As well as in the case of scattering polarized light by a dipole - oscillator in the direction of the perpendicular axes of a dipole, at $\delta = 0$, the intensity of scattered and falling non-polarized light is identical $I = I_0$.

3.3. Effective Section of Interaction of Electromagnetic Radiation with Coupled Atomic Electron

The problem of finding an effective section of scattering (reradiation) of linearly polarized light frequency ω in electronic orbit of an atom, figure 1.1.1, even in a classical variant, represents an issue of essential practical interest.

For the solution of this problem, it is necessary to take into account the existence of own frequency of oscillations of an electronic orbit of atom ω_0. The equation for oscillating electric dipole moment looks like the equation of the compelled oscillations (1.2.1).

The solution to the equation (1.2.1) appears thus (1.2.2): $\boldsymbol{p} = \dfrac{\dfrac{e^2}{m}\boldsymbol{E}_0 e^{i\omega t}}{\omega_0^2 - \omega^2 + i\gamma\omega}$. It is known that in a stationary orbit of the atom the electron does not radiate. But under the action of falling radiation of oscillations of an electronic orbit, this orbit has acceleration. There is an oscillating dipole which starts to radiate.

The flux of energy of radiation in a solid angle according to (3.1.19) is proportional to a square of acceleration of electronic orbit or to a square of the second derivative on time from the electric dipole moment:

$$dP = \frac{r_0^2}{\mu_0 c}\left[\frac{\partial^2 \mathrm{R}(\boldsymbol{p})}{\partial t^2}\right]^2 \sin^2\theta d\Omega =$$

$$= r_0^2 \varepsilon_0 c\left[\frac{\partial^2}{\partial t^2}\mathrm{R}\left(\frac{e^2}{m}\boldsymbol{E}_0 e^{i(\omega t-\Delta\varphi)}\frac{\cos(\Delta\varphi)}{\left(\omega_0^2-\omega^2\right)}\right)\right]^2 \sin^2\theta d\Omega = \quad (3.3.1)$$

$$= r_0^2 \varepsilon_0 c \frac{\omega^4\left(\frac{e^2}{m}\right)^2 E_0^2\left\langle\cos^2(\omega t-\Delta\varphi)\right\rangle}{\left(\omega_0^2-\omega^2\right)^2+\gamma^2\omega^2}\sin^2\theta d\Omega =$$

$$= r_0^2 \varepsilon_0 c \frac{\omega^4\left(\frac{e^2}{m}\right)^2 \frac{E_0^2}{2}}{\left(\omega_0^2-\omega^2\right)^2+\gamma^2\omega^2}\sin^2\theta d\Omega$$

where the real part of the dipole moment is used and averaging the square cosine on the period of oscillations of the electromagnetic field is executed.

In transformations it is convenient, on account of a phase delay (1.2.3), to use a ratio:

$$\omega_0^2-\omega^2+i\gamma\omega = \left(\omega_0^2-\omega^2\right)\left[1+i\mathrm{tg}(\Delta\varphi)\right] = \frac{\left(\omega_0^2-\omega^2\right)e^{i\Delta\varphi}}{\cos(\Delta\varphi)}.$$

At the analysis of the deduction there arises the question of whether it is possible in (3.3.1) to allocate the real part not in the dipole moment, but in a flux of energy of radiation. We have:

$$dP \sim \mathrm{R}\left[\frac{\partial^2 \boldsymbol{p}}{\partial t^2}\right]^2 = \mathrm{R}\left[\frac{\omega^4\left(\frac{e^2}{m}\right)^2 E_0^2 e^{2i(\omega t-\Delta\varphi)}}{\left(\omega_0^2-\omega^2\right)^2+\gamma^2\omega^2}\right] =.$$

$$= \frac{\omega^4\left(\frac{e^2}{m}\right)^2 E_0^2\left\langle\cos 2(\omega t-\Delta\varphi)\right\rangle}{\left(\omega_0^2-\omega^2\right)^2+\gamma^2\omega^2} = 0$$

The average for the period of the flux of energy of the radiated electromagnetic field in this case is equal to zero $dP = 0$. This result, obviously, contradicts physical sense since there is a situation when the first half of period energy is radiated by the oscillating electronic orbit, and the second half of the period is absorbed.

The given result is consequently that the complex value is not a flux of energy of radiation, and the electric dipole moment or its second derivative on time $\frac{\partial^2 \mathbf{p}}{\partial t^2} = e\mathbf{a}$, where is the acceleration of electronic orbit movement. Furthermore, as well as earlier, under the formula (3.1.4), using the average for the period intensity of falling radiation $I_0 = c\frac{\varepsilon_0 E_0^2}{2}$, we shall find the differential effective section of scattering of light on the oscillating electronic orbit of the atom:

$$d\sigma = \frac{dP}{I_0} = r_0^2 \varepsilon_0 c \frac{\omega^4 \left(\frac{e^2}{m}\right)^2 \frac{E_0^2}{2I_0}}{\left(\omega_0^2 - \omega^2\right)^2 + \gamma^2\omega^2} \sin^2\theta d\Omega = . \qquad (3.3.2)$$

$$= \frac{2\pi r_0^2 \omega^4}{\left(\omega_0^2 - \omega^2\right)^2 + \gamma^2\omega^2} \sin^3\theta d\theta$$

The intensity of reradiated light, according to (3.1.6), equals:

$$I = I_0 \frac{\omega^4}{\left(\omega_0^2 - \omega^2\right)^2 + \gamma^2\omega^2} \sin^2\theta \cdot \qquad (3.3.3)$$

Integrating, it is similar (3.2.5), the formula (3.3.2) on all space, we find the full effective section of scattering of polarized light:

$$\sigma = \frac{8\pi}{3} r_0^2 \frac{\omega^4}{\left(\omega_0^2 - \omega^2\right)^2 + \gamma^2\omega^2} . \qquad (3.3.4)$$

In the absence of attenuation and far from own frequency of an electronic orbit the formula (3.3.4) transforms into Thomson's formula (3.2.5). At rather small frequency of electromagnetic waves and in the absence of attenuation $\gamma = 0$, the effective section of scattering according to (3.3.4) is proportional to the fourth degree of frequency of radiation:

$$\sigma = \frac{8\pi}{3} r_0^2 \frac{\omega^4}{\omega_0^4} . \qquad (3.3.5)$$

The intensity of reradiation (scattering) in this case is:

$$I = I_0 \frac{\omega^4}{\omega_0^4} \sin^2\theta . \qquad (3.3.6)$$

The formula (3.3.6) reflects the so-called Rayleigh law of scattering $I \sim \omega^4$.

The size of the effective section of scattering grows at the approach of a resonance $\omega = \omega_0$.

In this case:

$$\sigma = \frac{8\pi}{3} r_0^2 \frac{\omega_0^2}{\gamma^2}. \tag{3.3.7}$$

At small attenuation the effective section of interaction of light with the atom can attain a very large size. It correlates with the infinite increase in the amplitude of the compelled oscillations in a resonance in the absence of attenuation.

The research executed in the given chapter on the basis of the classical analysis of the effective section of interaction of the electromagnetic radiation with atoms of substance shows that many physical phenomena - for example, the scattering of light by atoms - are rather informatively described by the given method.

Chapter 4

THE INTERACTION OF AN ELECTROMAGNETIC FIELD WITH A SUBSTANCE: THE QUANTUM APPROACH

In this chapter we transit to the quantum approach to the research of the interaction of the electromagnetic radiation with the substance.

Many physical parameters are quantized: energy of radiation, an electric charge, a magnetic flux, etc. In this line there is a physical problem, unresolved until now. It is a question regarding the elementary magnetic charge, the existence of which was noted by P. Dirac. Dirac's monopole - a magnetic charge - until now was not found out. There is an absence of elementary magnetic charge results in the asymmetry of Maxwell's equations concerning electric and magnetic fields. It is a disturbance of the aesthetics of the equations.

It is possible to approach the given problem as follows. There are no vetoes to consider a charge in the complexity value $e - i\mu$, where *e* is a charge of electron, - size of magnetic Dirac's monopole. The complexity of a charge defines, for example, the complexity of the constant detail structure:

$$\frac{(e - i\mu)^2}{hc} = \frac{e^2}{hc} + \frac{\mu^2}{hc} - 2i\frac{e\mu}{hc} = \frac{e^2}{hc} - \frac{\mu^2}{hc} + in = \alpha + \frac{n^2}{4\alpha} - in.$$

Thus, the theoretical size of a magnetic monopole follows a ratio $e\mu = \frac{1}{2}nhc$, where *n* as an integer (positive or negative) [60] is taken into account. At we have a complex constant of the detailed structure as $\frac{(e - i\mu)^2}{hc} \approx \frac{1}{137} + \frac{137}{4} - i$.

Thus, Maxwell's equations can be written down in a more symmetrical form $\frac{\partial \boldsymbol{B}}{\partial t} - \boldsymbol{j}_m = -\,\mathrm{rot}\boldsymbol{E}$, $\frac{\partial \boldsymbol{D}}{\partial t} + \boldsymbol{j} = \mathrm{rot}\boldsymbol{H}$, and $\mathrm{div}\,\boldsymbol{B} = -\rho_m$, where and are the density of a current and the volumetric density of free magnetic charges - imaginary sizes. Usually these equations are written down in the real area, so and $\rho_m = 0$.

From a practical point of view, we can note that the quantum approach is correct for short lengths of waves. Already for MW-radiation with the length of a wave more than 10 *mm* and, furthermore, in the areas of meter and kilometer waves, the quantum approach is not adequate.

As well as in the case of the classical approach, in the quantum approach there is a line of methods of research of interaction of electromagnetic radiation with the substance. Among them we shall note a method of the impacts of quantums (or photons) with the particles, a method of the kinetic equations, and, certainly, a method of effective sections which in the quantum area have a very big information force.

4.1. Research of Interaction of Quantums with Particles of Substance by a Method of Impacts: A Photoeffect

The interaction of electromagnetic radiation with atoms can cause their ionization. The occurrence of free electrons under the action of electromagnetic radiation (ionizing radiations) is referred to as a photoeffect.

The photoeffect is a process of absorption of a photon coupled with an electron, when the energy of a photon exceeds its energy of coupling or work output ($h\nu > A_{\text{out}}$) where is Planck's constant, ν - frequency of quantum. The interaction of photons of high energy occurs to the electrons which takes place on the internal shells of atoms. Electrons are torn away from the atoms.

The energy of a photon is spent for work of an output of electron from atom and on the transfer to the electron of kinetic energy $\frac{m_1 V^2}{2}$, where is the mass of the electron, V – the velocity, figure 4.1.1. The law of energy conservation looks like Einstein's formula:

$$h\nu = A_{\text{out}} + \frac{m_1 V^2}{2}.$$

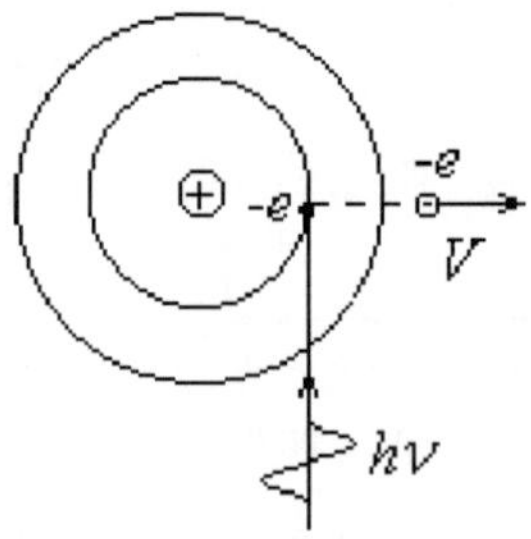

Figure 4.1.1.

The impulse of quantum is transferred through the internal orbital electron strongly coupled to a nucleus. At an irradiation of a surface of metal interaction of radiation occurs basically to the top layers of atoms. But the angular distribution of a start of photoelectrons from individual atoms, i.e. at an internal photoeffect, will be appreciably reflected in the distribution of a start of these electrons from a metal surface, i.e. in an external photoeffect.

Therefore, it is of interest to consider the angular distribution of photoelectrons at an internal photoeffect.

4.1.1. Not Relativistic Case

In spite of the fact that the nature of a photoeffect has been explained by Einstein at the beginning of the twentieth century, various aspects of this phenomenon still draw attention. So, in the description of the angular distribution of the photoelectrons which are torn out by photons from atoms, there are significant disagreements. For example, in [22], using a method of calculation of Feynman diagrams, we can conclude that the photoelectrons start forward in the direction of photon movement and back in approximation of the main order at the photoeffect is absent.

In [4] it is also underlined that in the direction of propagation of quantum the photoelectrons do not fly out. This conclusion is made on the basis of positions which in the simplified variant are represented by the following. The impulse, the started electron, is defined basically by its action on the electron, an electric vector of light quantum. In case the electron starts in the direction of quantum electric vector, it gets an impulse p_e.

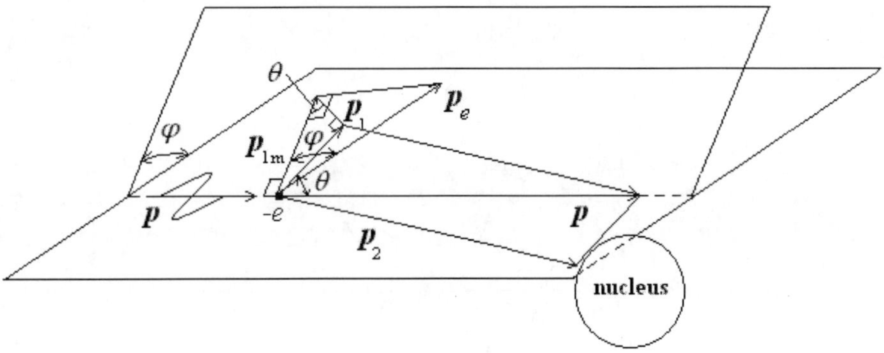

Figure 4.1.2.

In a plane located under angle φ to a polarization plane of light quantum, figure 4.1.2, the size of the electron impulse $p_{1m} = p_e \cos\varphi$. If the electron impulse besides it is directed under angle θ to the direction of light quantum its size will make value:

$$p_1 = p_e \cos\varphi \sin\theta . \tag{4.1.1}$$

In figure 4.1.2 two equal angles θ with mutually perpendicular sides are marked. Hence, the energy of a photoelectron equals:

$$E_1 = \frac{p_1^2}{2m_1} = \frac{p_e^2 \cos^2\varphi \sin^2\theta}{2m_1}, \quad (4.1.2)$$

where is the mass of electron.

At the energy of photoelectrons is $E_1 = 0$. Photoelectrons in the maximum start in a direction of a light vector or vector of polarization, i.e. a vector of electric field strength of light quantum. The same dependence is offered and in [19].

The formula (4.1.2) has the simplified character in comparison [4, 19], but correctly reflects the basic dependence of energy of distribution of a start of photoelectrons on angles φ and θ.

Dependence (4.1.2) has something lacking; which is that at its deduction the law of conservation of impulse was not used; therefore, there is no electron movement in the direction $\theta = 0$. Using the conservation equation of impulse in [4, 19] cannot be considered satisfactory since in the analysis carried out by the authors it has an auxiliary character. In the analysis, the transition of the electron from a discrete energy spectrum in a state of a continuous spectrum under the influence of harmonious perturbation lays, i.e. the matrix element of the operator of perturbation is the harmonious function of time. In other words, the attention is paid to the wave nature of the quantum interacting with electron. Angular distribution of electron energy in the relative units, plotted under the formula (4.1.2), is shown in figure 4.1.3, a curve 1.

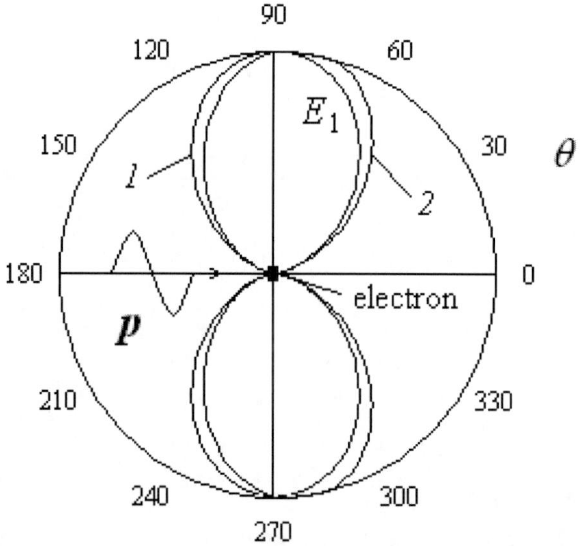

Figure 4.1.3.

Let's illustrate the occurrence of the amendment to the formula (4.1.2) connected to the existence of the impulse of a photon $\boldsymbol{p}$, following [23]. In figure 4.1.4 the change of the photoelectron energy is shown at the existence of a photon impulse.

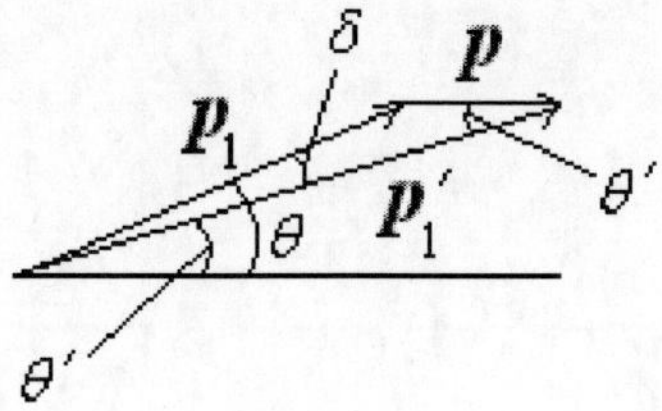

Figure 4.1.4.

From the figure, it follows that $\theta = \theta' + \delta$. We shall find $\sin\theta = \sin\theta' \cos\delta + \sin\delta\cos\theta'$. Believing δ is small, we have:

$$\sin\theta = \sin\theta'\left(1 + \frac{\sin\delta\cos\theta'}{\sin\theta'}\right) = \sin\theta'\left(1 + \frac{p}{p_1}\cos\theta'\right).$$

Thus, the theorem of sine for a triangle in figure 4.1.4 is used. Furthermore, using Einstein's formula, and considering where is the ratio of a photoelectron velocity to the light velocity in vacuum, we have $\sin\theta = \sin\theta'\left(1 + \beta'\cos\theta'\right)$. Taking into account a little sized β', the formula (4.1.2) we shall transform to a form $E_1 = \dfrac{p_e^2\cos^2\varphi\sin^2\theta'}{2m_1}\left(1 + 2\beta'\cos\theta'\right)$. The angular distribution of electrons' energy for plotted under the formula (4.1.2) in view of the amendment is shown in figure 4.1.3, a curve 2. Thus, the indicator of photoelectrons' scattering has received some inclination forward, but in a direction of a quantum impulse, i.e. at $\theta = 0$, the electrons, as before, do not start.

The formula (4.1.2) has an explanation on the basis of the wave nature of light. For the proof of this position we shall consider the interaction of the electromagnetic wave with the orbital electron. The description of the orbital electron movement, figure 4.1.5, is carried out on the basis of Bohr's semi-classical theories since the process of interaction of the electron with the electromagnetic wave is researched from the position of classical physics.

From triangles of speeds, we shall find:

$$\sin\alpha = \frac{V_1}{V_t}\cos\theta, \qquad (4.1.3)$$

where is the velocity of the electron movement around of a nucleus, - total velocity of the electron in view of the action on it of an electromagnetic wave.

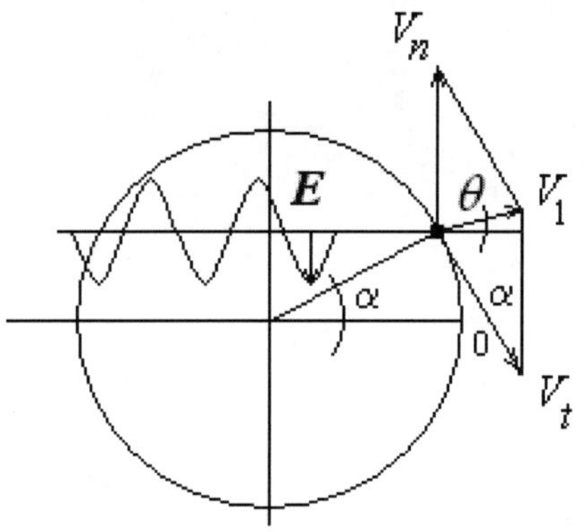

Figure 4.1.5.

Under the theorem of cosines we have:

$$V_1^2 = V_n^2 + V_t^2 - 2V_nV_t\cos\alpha = V_n^2 + V_t^2 - 2V_nV_t\sqrt{1-\left(\frac{V_1}{V_t}\cos\theta\right)^2}, \tag{4.1.4}$$

where is a component of the general velocity of the electron movement after its outlet from a nucleus which arises under the action of electric field strength in an electromagnetic wave.

Solving (4.1.4) relatively V_1, we shall find:

$$V_1^2 = \left(V_n^2 + V_t^2 - 2V_n^2\cos^2\theta\right) + \sqrt{\left(V_n^2 + V_t^2 - 2V_n^2\cos^2\theta\right)^2 - \left(V_n^2 - V_t^2\right)^2}. \tag{4.1.5}$$

The condition of a tear of electrons away from the atom at any electron position is $V_n \geq V_t$. In the case of equality of velocities , we have:

$$V_1 = 2V_n \sin\theta. \tag{4.1.6}$$

The distribution of velocities (4.1.6) corresponds (4.1.2) and figure 4.1.3, a curve 1. Thus, the ratio (4.1.6) arises if we take into account only the wave nature of the electromagnetic wave which interacts with the orbital electron.

In [18] the start of the distribution of an angle of electrons it is also investigated for a relativistic case. Thus, the electrons are outlet mainly in the direction of the propagation of a photon. However, the deduction carried out is actually based on the formula (4.1.1). Lack of deduction [18] shows the absence of the final formulas of angular electron distribution the mass of a nucleus m_2.

Moreover, in fact, the mass of a nucleus defines a portion of photon impulse which can receive a nucleus. Let's consider the phenomenon of an internal photoeffect from the positions of corpuscular representation of light quantum, figure 4.1.2. The light quantum by the impulse and energy E tears the electron away from the atom, making work of an output A_{out}. Thus, it should be observed as the law of energy conservation:

$$E = A_{out} + E_1 + E_2, \tag{4.1.7}$$

where is kinetic energy let out of the electron, - kinetic energy of a nucleus, and the law of conservation of impulse:

$$\boldsymbol{p} = \boldsymbol{p}_1 + \boldsymbol{p}_2, \tag{4.1.8}$$

where is the impulse let out of the electron, - impulse transferred to a nucleus, figure 4.1.2.

The formula (4.1.7) differs from Einstein's standard formula a little. The issue is that Einstein's formula means an absence of angular distribution of the photoelectrons' velocity. Really, if the energy of photon E and work of output for the given chemical element is set, thus, a certain velocity of a start electron from the atom is determined. This means that the velocities of electrons, let out for every possible direction, are identical, and the task of finding their angular distribution becomes incorrect. Such a position is a result of the absence in Einstein's formula of the kinetic energy of a nucleus E_2.

The size of the impulse transferred to a nucleus is possible to find under the formula following from (4.1.8):

$$p_2^2 = p^2 + p_1^2 - 2pp_1\cos\theta. \tag{4.1.9}$$

For the common solving of the system of equations (4.1.7) and (4.1.9), the equation (4.1.9) is convenient for presenting through energy. Taking into account , where c is light velocity in vacuum, and $p_2^2 = 2m_2E_2$, we shall find:

$$2m_2E_2 = \left(\frac{E}{c}\right)^2 + 2m_1E_1 - 2\frac{E}{c}\sqrt{2m_1E_1}\cos\theta, \tag{4.1.10}$$

where is the mass of the electron, - mass of a nucleus.

Substituting in (4.1.10) the kinetic energy of a nucleus from (4.1.7), we have:

$$E - A_{out} - E_1 = \frac{1}{2m_2}\left(\frac{E}{c}\right)^2 + \frac{m_1}{m_2}E_1 - \frac{E}{m_2c}\sqrt{2m_1E_1}\cos\theta \cdot \tag{4.1.11}$$

We use designations $G = \sqrt{E_1}$, $\alpha = \frac{E}{m_2c}\sqrt{2m_1}$, $\sigma = 1 + \frac{m_1}{m_2}$, $\gamma = \frac{1}{2m_2}\left(\frac{E}{c}\right)^2 + A_{out} - E \cdot$

Then the equation (4.1.11) will be transformed to this kind:

$$\beta G^2 - \alpha G\cos\theta + \gamma = 0. \qquad (4.1.12)$$

Solving a quadratic (4.1.12) under the condition of (the mass of the electron is much less than the mass of a nucleus), we find:

$$G_{1,2} = \frac{\alpha}{2}\cos\theta \pm \sqrt{\frac{\alpha^2}{4}\cos^2\theta - \gamma}\,. \qquad (4.1.13)$$

Substituting in (4.1.13) the accepted designations, we have:

$$G_{1,2} = \sqrt{\frac{m_1}{2}}\frac{E}{m_2 c}\cos\theta \pm \sqrt{\frac{E^2}{2m_2c^2}\left(\frac{m_1}{m_2}\cos^2\theta - 1\right) + E - A_{out}}\,. \qquad (4.1.14)$$

Taking into account that , where is the velocity of photoelectrons, and also the condition $\frac{m_1}{m_2}\cos^2\theta << 1$, we shall find:

$$V_1 = \frac{E}{m_2 c}\cos\theta \pm \sqrt{\left(\frac{m_1}{m_2}\cos^2\theta - 1\right)\frac{m_2}{m_1}\left(\frac{E}{m_2 c}\right)^2 + \frac{2(E - A_{out})}{m_1}}\,. \qquad (4.1.15)$$

At a tendency of a nuclear mass to infinity $m_2 \to \infty$, the formula (4.1.15) transforms in Einstein's standard law for a photoeffect. Besides, as it was already specified earlier, there is no angular distribution of photoelectrons' velocity.

The condition is correct at the external photoeffect when the impulse of a photon is transferred through separate atoms to all of the metal as a whole. Therefore, for an external photoeffect, i.e. for the interaction of a solid body and a photon, Einstein's formula is absolutely applicable.

For an internal photoeffect in the formula (4.1.15), it is necessary to use the effective mass of a nucleus which takes into account the forces of an attraction between atoms in a substance.

Transforming (4.1.15), we shall obtain:

$$V_1 = \frac{E}{m_2 c}\left(\cos\theta \pm \sqrt{\cos^2\theta + \frac{m_2}{m_1}\left(2(E - A_{out})\frac{m_2 c^2}{E^2} - 1\right)}\right). \qquad (4.1.16)$$

Let's designate $\Delta = E - A_{out}$. The angular distribution of photoelectrons will arise at $\Delta \geq \frac{E^2}{2m_2c^2}$. In the right part of the received inequality there is a very small size; therefore, the distribution of photoelectrons will appear practically at $E > A_{out}$.

Let's designate $\Delta = \eta \frac{E^2}{2m_2c^2}$, where characterizes the size of excess of the photon energy above the work of output in relative units. Thus, the formula (4.1.16) takes the form:

$$V_1 = \frac{E}{m_2 c}\left(\cos\theta + \sqrt{\cos^2\theta + \frac{m_2}{m_1}(\eta - 1)}\right). \tag{4.1.17}$$

The analysis of the formula (4.1.17) shows that before a root it is necessary to take a sign plus since differently scattering electrons basically occurs back to the direction of a falling photon.

In figure 4.1.6 the angular distribution of a start of electrons is shown at internal photoeffect in the relative units $\frac{V_1}{E/m_2 c}$, plotted under the formula (4.1.17) at several values η for copper.

From the figure it is visible that the velocity of photoelectrons becomes practically identical in all directions already at $\eta \geq 1.01$. Then, Einstein's formula is correct for the internal photoeffect. Taking into account that, for example, for copper in the field of a red border of a photoeffect ($\lambda_r = 250\ nm$) it is possible to judge the ratio, under the order of size equivalent $\frac{E^2}{2m_2c^2} \Big/ \Delta = \frac{1}{\eta}$, that the difference of distribution of photoelectrons velocities from spherical is visible, i.e. actual infringement of the formula can be observed only in a very short-wave area of the spectrum of γ- radiations.

In figure 4.1.6 black squares show the results of the measurement of the angular distribution of photoelectrons, torn off by quantums from a monolayer of the copper atoms; the nickel put on a surface [24].

The length of a wave of quantums allows us to observe a photoeffect c $2p$ - shells of copper atoms, but the photoeffect on nickel thus was absent.

№	η
1	1
2	1.00001
3	1.0001
4	1.01

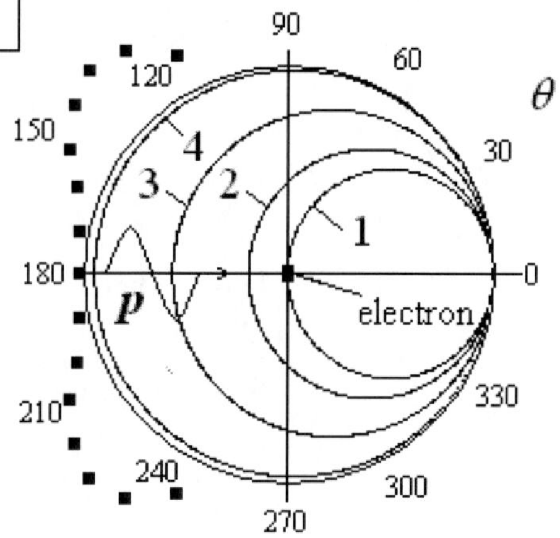

Figure 4.1.6.

The experimental distribution of photoelectrons contradicts with calculating the distribution to figure 4.1.3. Moreover, as opposed to figure 4.1.3, small maximums of the indicator of the distributions directed opposite to a direction of flight of light quantums under an angle of approximately to a direction of a light flux are observed. In [24] these maximums explain the focusing properties of all totality of atoms of a surface. At an increase in quantity of the monolayers of atoms of copper at nickel, the amplitude of maximums grows.

Thus, depending on what properties - wave or corpuscular - the light quantum shows at the interaction with orbital electron, the angular distribution of photoelectrons will be completely various. The answer to the question of what distribution is correct, figure 4.1.3 or figure 4.1.6, can be found only by experiment. However, the existence of electron flux from an irradiated surface at normal falling light [24], in a direction opposite to the intensity of light specifies the prevalence of the corpuscular properties of light at its interaction with atoms.

4.1.2. Relativistic Case

By consideration of a relativistic variant of an internal photoeffect the law of energy conservation needs to be written down as:

$$E = A_{\text{out}} + E_{\text{kin}} + E_2 , \tag{4.1.18}$$

where is the kinetic energy of a photoelectron.

The law of impulse conservation remains in the form (4.1.9). Using a relativistic connection between the energy and impulse for electron:

$$E_1^2 = p_1^2 c^2 + m_1^2 c^4, \tag{4.1.19}$$

where is full electron energy, - in this case its mass of the rest, we shall deduce from (4.1.19) the electron impulse and we shall substitute it in the equation (4.1.9). For convenience of the further transformations we shall write down (4.1.19) as:

$$p_1^2 = \frac{E_1^2 - m_1^2 c^4}{c^2} = \frac{\left(E_1 + m_1 c^2\right) E_{\text{kin}}}{c^2}. \tag{4.1.20}$$

At the writing (4.1.20) the ratio is used:

$$E_{\text{kin}} = E_1 - m_1 c^2. \tag{4.1.21}$$

The equation (4.1.9) will be transformed to a form:

$$2m_2 c^2 E_2 = E^2 + \left(E_1 + m_1 c^2\right) E_{\text{kin}} - 2E\sqrt{\left(E_1 + m_1 c^2\right) E_{\text{kin}}} \cos\theta. \tag{4.1.22}$$

In connection with the big mass of a nucleus and its relative low velocity after interaction with a photon, the expression for the connection of a nucleus impulse with its kinetic energy is not used in a relativistic form.

Having substituted in (4.1.22) size from (4.1.18), we obtain:

$$\begin{aligned} &2m_2 c^2 \left(E - A_{\text{out}} - E_{\text{kin}}\right) = \\ &= E^2 + \left(E_1 + m_1 c^2\right) E_{\text{kin}} - 2E\sqrt{\left(E_1 + m_1 c^2\right) E_{\text{kin}}} \cos\theta \end{aligned} \tag{4.1.23}$$

Let's designate:

$$\alpha = \frac{E\sqrt{E_1 + m_1 c^2}}{m_2 c^2}; \qquad G = \sqrt{E_{\text{kin}}}. \tag{4.1.24}$$

As a result, it will be transformed (4.1.23):

$$\delta G^2 - \alpha G \cos\theta + \gamma = 0. \tag{4.1.25}$$

The designation

$$\gamma = \frac{1}{2m_2}\left(\frac{E}{c}\right)^2 + A_{\text{out}} - E = \frac{1}{2m_2}\left(\frac{E}{c}\right)^2 - \Delta = \frac{1}{2m_2}\left(\frac{E}{c}\right)^2 (1 - \eta)$$

corresponds (4.1.12). The value

$$\delta = 1 + \frac{m_2 c^2}{2}\left(\frac{\alpha}{E}\right)^2 = 1 + \frac{m_1}{2m_2} + \frac{E_1}{2m_2 c^2} = 1 + \frac{m_1}{2m_2} + \frac{E_{\text{kin}} - m_1 c^2}{2m_2 c^2} = 1 + \frac{E_{\text{kin}}}{2m_2 c^2} \approx 1 \cdot$$

Thus it is taken into account that $E_{\text{kin}} << 2m_2 c^2$.

Solving the equation (4.1.25), we obtain:

$$G_{1,2} = \frac{\alpha}{2}\cos\theta \pm \sqrt{\frac{\alpha^2}{4}\cos^2\theta - \gamma} \, . \qquad (4.1.26)$$

Substituting designations, we find:

$$E_{\text{кин}} = \left(\frac{\alpha}{2}\right)^2 \left(\cos\theta + \sqrt{\cos^2\theta + \frac{2}{\alpha^2}\left(\frac{E^2}{m_2 c^2}\right)(\eta - 1)}\right)^2 \cdot \qquad (4.1.27)$$

As opposed to the non-relativistic case, the formula (4.1.17), in the formula (4.1.27) in the right part there is a value which depends on the full electron energy , the structure of which also includes kinetic energy E_{kin}. But dependence of size α from is not as strong as the structure of full energy which includes rather larger electron energy of rest $m_1 c^2$.

Taking into account , where the relative velocity of a photoelectron, we shall find:

$$\frac{\alpha^2}{2} = \left(\frac{E}{m_2 c}\right)^2 \frac{m_1}{2}\left(\frac{1}{\sqrt{1-\beta^2}} + 1\right). \qquad (4.1.28)$$

Substituting (4.1.28) in (4.1.27) and taking into account $E_{\text{kin}} = m_1 c^2 \left(\frac{1}{\sqrt{1-\beta^2}} - 1\right)$, we shall obtain:

$$2c\sqrt{(1-\mu)} = \frac{E}{m_2 c}\left(\cos\theta + \sqrt{\cos^2\theta + \mu\frac{m_2}{m_1}(\eta - 1)}\right), \qquad (4.1.29)$$

where it is designated $\mu = \dfrac{2}{\dfrac{1}{\sqrt{1-\beta^2}}+1}$.

Taking into account $2c\sqrt{(1-\mu)} = \sqrt{\dfrac{2E_{\text{kin}}}{m_1} \Big/ \left(1+\dfrac{E_{\text{kin}}}{2m_1c^2}\right)} \approx V_1$,

at $1 >> \dfrac{E_{\text{kin}}}{2m_1c^2}$, we shall find:

$$V_1 \approx \frac{E}{m_2c}\left(\cos\theta + \sqrt{\cos^2\theta + \mu\frac{m_2}{m_1}(\eta - 1)}\right). \tag{4.1.30}$$

The formula (4.1.30) allows us to take into account the relativistic effect at a photoeffect, in the case of high velocities of photoelectrons. Thus, as opposed to (4.1.17), under a root relativistic factor μ is used. As the calculation of dependence is shown on figure 4.1.7, until the velocities of the photoelectrons equal about half of the light velocity in a vacuum by the relativistic effects at the calculation of the distribution of a start of photoelectrons, it is possible to neglect and use the formula (4.1.17).

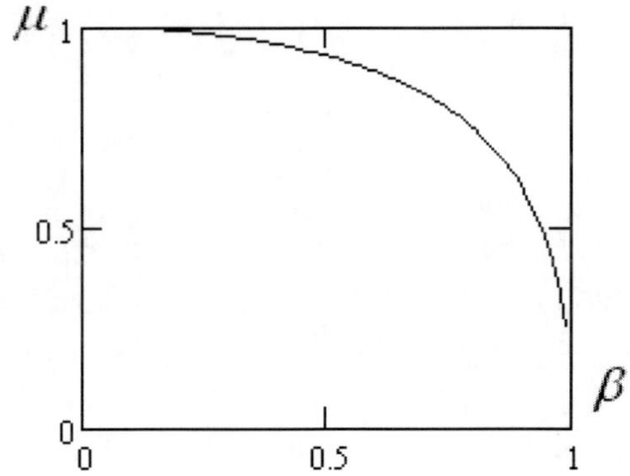

Figure 4.1.7.

4.2. Interaction of a Photon with Loosely Coupled or Free Electrons

Compton-Effect

Interaction of photons with free or loosely coupled electrons in substance represents known enough problem. One of results of such interaction, for example, is not coherent scattering. Effect of Compton (Compton-effect) or not coherent scattering of X-rays on the electrons, loosely coupled with a nucleus, at an irradiation of a paraffin surface which was observed for the first time by A. H. Compton, is well investigated problem [25]. In a spectrum of scattered radiation he observed lower frequency, than frequency of falling radiation. Effect of Compton represents a vivid example of display of the quantum nature of light since there is no classical analogue of this phenomenon which could be described Maxwell equations.

The Compton-effect is an interaction of high energy photons with loosely coupled electrons external orbits of atom (or with free electrons).

With some approximation, the law of energy conservation at Compton-effect looks like:

$$h\nu = h\nu' + A_{\text{out}} + \frac{mV^2}{2}.$$

There is a tear of electrons away from the atom with the appearance of recoil electron, figure 4.2.1.

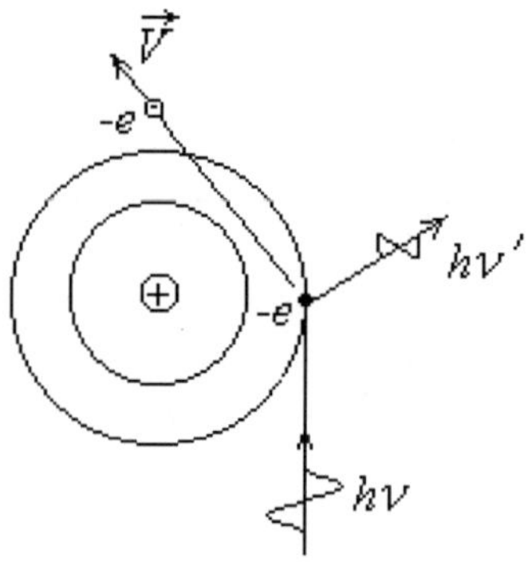

Figure 4.2.1.

The energy of a primary photon is great ($h\nu > A_{\text{out}}$) and its impulse to give a nucleus is impossible, since the external orbital electron is loosely coupled to a nucleus. Therefore, there is the secondary quantum which takes almost all of the impulse. The Compton-effect occurs with both the absorption of radiation energy and the scattering of energy - quantum $h\nu'$. The frequency of secondary radiation is lower than the primary ν since the energy of quantum of secondary radiation is less.

According to the Compton-effect, as well as the photoeffect, there occurs a tear of electron away from the atom. Therefore, these radiations are ionizing radiations.

There have been attempts to describe the effect of Compton on the basis of wave effect of Doppler [26]. Thus, the electron is again supposed as the mobile receiver of a wave, and then as a mobile source of a wave. However, there are still similar attempts which Compton himself criticized [61]. We can note that the scattering on a motionless electron, which in a strict mathematical sense is considered to be the effect of Compton, in general is not entered in any way in the given model.

A rather detailed analysis of the Compton-effect is present in [18]. But this analysis is carried out in the assumption of the absence of an influence from the surrounding substance on the process.

Our purpose will be the analysis, first of all, not the coherent shift of photon frequency and possible outcomes, at the interaction of a photon and a free electron in substance, and also the influence of the characteristics of the substance on the examined processes.

Research of the noncoherent shift of frequency finally defines a direction (the diagram of distribution) of impulses of recoil electron movement.

4.2.1. Laws of Interaction of a Photon with a Free Electron

Let's consider the interaction of a moving electron with a photon, as shown in figure 4.2.2. In all further transformations, we believe that the reference system is connected to motionless units of a crystal lattice of substance.

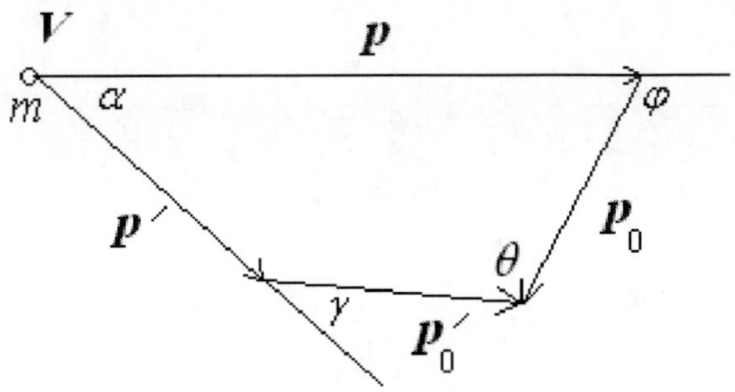

Figure 4.2.2.

We shall write down the law of energy conservation at the interaction of a photon with a moving electron as:

$$E + h\nu = E^{/} + h\nu^{/}, \qquad (4.2.1)$$

where E and are electron energy before and after interaction with a photon, and ν and - are the photon frequency before and after collision with electron.

Impulse of a photon before interaction with electron $p_0 = \dfrac{h\nu}{V_{ph}} = \dfrac{nh\nu}{c}$, where is velocity of a photon in substance, - an absolute parameter of substance refraction, c - velocity of light in vacuum. Hence, the formula (4.2.1) will be copied as:

$$E + \frac{p_0 c}{n} = E^{/} + \frac{p_0^{/} c}{n}, \qquad (4.2.2)$$

where is a photon impulse after interaction with electron.

The law of impulse conservation before and after interaction, figure 4.2.2, looks like:

$$\boldsymbol{p} + \boldsymbol{p}_0 = \boldsymbol{p}' + \boldsymbol{p}_0', \tag{4.2.3}$$

Let's transform (4.2.3) to the scalar form:

$$p_0^2 + p_0'^2 - 2p_0p_0'\cos\theta = p^2 + p'^2 - 2pp'\cos\alpha \tag{4.2.4}$$

where α is the angle between the directions of electron impulse before and after interaction with a photon, and θ is the angle between impulses of falling and scattered photons. Using the connection between the energy and velocity of the electron where $p = mV$, values m and V - mass and velocity of electron before collision, we shall transform the equation (4.2.1) to a form:

$$\frac{pc}{V} + \frac{p_0}{n} = \frac{p'c}{V'} + \frac{p_0'}{n}. \tag{4.2.5}$$

In the formula (4.2.5) $E' = \dfrac{p'}{V'}c^2$, and is the velocity of theelectron after collision with a photon.

Let's exclude from (4.2.4) and (4.2.5) the value p'. Finding from (4.2.5) the value and substituting it in (4.2.4), we shall obtain:

$$p_0^2 + p_0'^2 - 2p_0p_0'\cos\theta = p^2 + \left[\frac{V'}{c}\left(p\frac{c}{V} + \frac{p_0}{n} - \frac{p_0'}{n}\right)\right]^2 - \tag{4.2.6}$$
$$-2p\frac{V'}{c}\left(p\frac{c}{V} + \frac{p_0}{n} - \frac{p_0'}{n}\right)\cos\alpha$$

Let's exclude from (4.2.6) the electron velocity after collision with a photon. By determination of the particle impulse we have:

$$p' = \frac{m_0V'}{\sqrt{1 - \dfrac{V'^2}{c^2}}}, \tag{4.2.7}$$

where is an electron rest mass.

Substituting (4.2.7) in (4.2.5), we shall find:

$$\frac{m_0 c}{\sqrt{1-\frac{V'^2}{c^2}}}=\frac{pc}{V}+\frac{p_0}{n}-\frac{p_0'}{n}. \tag{4.2.8}$$

Finding the solution (4.2.8) relative V', we have:

$$\frac{V'}{c}=\sqrt{1-\left(\frac{m_0 c}{\frac{pc}{V}+\frac{p_0}{n}-\frac{p_0'}{n}}\right)^2}. \tag{4.2.9}$$

Substituting (4.2.9) in (4.2.6), we obtain:

$$p_0^2+p_0'^2-2p_0p_0'\cos\theta=p^2+\left(p\frac{c}{V}+\frac{p_0}{n}-\frac{p_0'}{n}\right)^2-(m_0c)^2-$$
$$-2p\left(\sqrt{\left(p\frac{c}{V}+\frac{p_0}{n}-\frac{p_0'}{n}\right)^2-(m_0c)^2}\right)\cos\alpha. \tag{4.2.10}$$

The law of impulse conservation (4.2.3) in the scalar form can be written down differently (4.2.4):

$$p^2+p_0^2-2pp_0\cos\left(180^o-\varphi\right)=p'^2+p_0'^2-2p'p_0'\cos\left(180^o-\gamma\right), \tag{4.2.11}$$

or:

$$p^2+p_0^2+2pp_0\cos\varphi=p'^2+p_0'^2+2p'p_0'\cos\gamma, \tag{4.2.12}$$

where φ is the angle between directions of movement of the electron and photon before collision, γ - the angle between directions of the electron and photon after collision.

As well as in the previous case, substituting value from (4.2.5) in (4.2.12), we have:

$$p^2+p_0^2+2pp_0\cos\varphi=$$
$$=\left(\frac{V'}{c}\right)^2\left(p\frac{c}{V}+\frac{p_0}{n}-\frac{p_0'}{n}\right)^2+p_0'^2+2p_0'\left(\frac{V'}{c}\right)\left(p\frac{c}{V}+\frac{p_0}{n}-\frac{p_0'}{n}\right)\cos\gamma. \tag{4.2.13}$$

Furthermore, using (4.2.9), we shall copy (4.2.13) as:

$$p^2 + p_0^2 + 2pp_0\cos\varphi = \left(p\frac{c}{V} + \frac{p_0}{n} - \frac{p_0^/}{n}\right)^2 - (m_0c)^2 + p_0^{/2} + \\ + 2p_0^/\left(\sqrt{\left(p\frac{c}{V} + \frac{p_0}{n} - \frac{p_0^/}{n}\right)^2 - (m_0c)^2}\right)\cos\gamma \quad (4.2.14)$$

Let's consider some special cases of transformation (4.2.10) and (4.2.14) which are not excessively cumbersome, and which will allow us to carry out the analysis of results obviously enough.

4.2.2. Absorption of Photon by a Free Electron Moving in Substance

Let's assume that a photon falling on a free-moving electron is completely absorbed. Thus, the condition should satisfy. From (4.2.14) we find:

$$p^2 + p_0^2 + 2pp_0\cos\varphi = \left(p\frac{c}{V} + \frac{p_0}{n}\right)^2 - (m_0c)^2. \quad (4.2.15)$$

Carrying out transformations, we shall receive:

$$p^2\left[1 - \left(\frac{c}{V}\right)^2\right] + p_0^2\left(1 - \frac{1}{n^2}\right) = 2pp_0\left(\frac{c}{nV} - \cos\varphi\right) - (m_0c)^2. \quad (4.2.16)$$

Using a ratio $p = m_0V \Big/ \sqrt{1 - \frac{V^2}{c^2}}$, we have:

$$\frac{(m_0V)^2}{\left(1 - \frac{V^2}{c^2}\right)}\left[1 - \left(\frac{c}{V}\right)^2\right] + p_0^2\left(1 - \frac{1}{n^2}\right) = 2pp_0\left(\frac{c}{nV} - \cos\varphi\right) - (m_0c)^2. \quad (4.2.17)$$

It is simple to see that the first addend in the left part (4.2.17) is equal $-(m_0c)^2$. Hence:

$$\cos\varphi = \frac{c}{nV} - \frac{p_0}{2p}\left(1 - \frac{1}{n^2}\right). \quad (4.2.18)$$

The received expression is an analogue of the formula of the Vavilov-Cherenkov effect [21] for the angle of the radiated photon, by uniformly moving the electron in media. The difference is only in a minus sign in the right part (4.2.18) which is connected to a change of direction of a photon impulse p_0.

From the formula (4.2.18), it follows that in vacuum, i.e. at $n=1$, the free electron cannot absorb the photon, since $\cos\varphi > 1$. But in substance, at condition:

$$\frac{c}{nV} - \frac{p_0}{2p}\left(1-\frac{1}{n^2}\right) \le 1, \tag{4.2.19}$$

such a process is possible.

Let's consider in more detail a condition of photon absorption by the electron. We shall transform:

$$\frac{p_0}{p} = \frac{nh\nu}{mVc} \approx \frac{nh\nu}{m_0Vc} = \frac{n\Lambda\nu}{V} = \frac{n\Lambda c}{Vn\lambda} = \frac{\Lambda}{\lambda}\frac{c}{V}, \tag{4.2.20}$$

where size is Compton's length of electron wave. It is accepted also that the velocity of the electron is much less than the light velocity in vacuum, so $m \approx m_0$. Hence, the formula (4.2.19) becomes:

$$\frac{\Lambda}{\lambda}\left(1-\frac{1}{n^2}\right) \ge \frac{2}{n}. \tag{4.2.21}$$

For the inequalities (4.2.19) and (4.2.21) are not corrected. If a parameter of refraction is $n>1$, we obtain:

$$\lambda \le \frac{\Lambda}{2}\left(n-\frac{1}{n}\right). \tag{4.2.22}$$

The formula (4.2.22) shows that the absorption of a photon by a free electron, moving in substance with a parameter of refraction of is possible, apparently, only in the area of X-ray radiation. Moreover, the greater the parameter of refraction, the bigger the length of a wave that can absorb the electron.

For lines of the substances considered in [23], the parameter of refraction in the area of X-ray radiation is less than one, and it is very close to one. The vacuum for X-ray radiation is, in comparison with these substances, a denser optical medium. For them, the absorption of photons by an electron according to (4.2.22) is impossible. However, there is no physical basis to completely deny the existence of substances with a parameter of refraction more in the area of X-ray radiation. Free electrons in such substances could absorb quantums of X-ray radiation.

4.2.3. Scattering the Photon on Free Electron

a) Coherent Scattering

Coherent scattering of quantum can be observed not only on electron strongly coupled to a nucleus, but also on free electron.

Coherent scattering of a photon on free-moving electron occurs if the condition is observed. The formula (4.2.5) will be transformed to: $p = p^{/}$, i.e. the electron changes only in the direction of the movement. In this case, the equation (4.2.4) gives:

$$p_0 \sin\frac{\theta}{2} = p \sin\frac{\alpha}{2}. \tag{4.2.23}$$

The equation (4.2.12) results in a condition:

$$\cos\varphi = \cos\gamma. \tag{4.2.24}$$

The condition (4.2.24) shows a coherent scattering $\varphi = \gamma$. The given condition is actually the law of reflection of the photon, similar to the law of reflection of light from the mirror interfaces of two media.

Formulas (4.2.23) and (4.2.24) also show that at a coherent scattering the quadrangle of impulses on figure 4.2.2 becomes symmetrical, see figure 4.2.3.

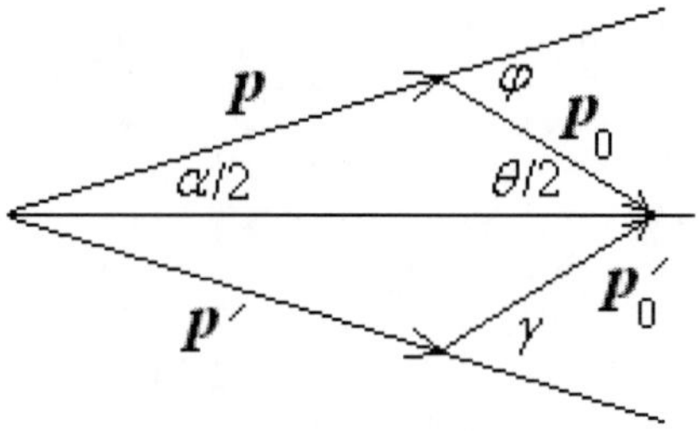

Figure 4.2.3.

b) Compton-Effect

Parameters of the Compton-effect are found, provided that the electron is motionless [21], i.e. $p = 0$, $V = 0$, $\frac{pc}{V} = m_0 c \Big/ \sqrt{1 - \frac{V^2}{c^2}} = m_0 c$.

In this case, the formula (4.2.10) becomes:

$$p_0^{\ 2} + p_0^{/2} - 2p_0 p_0^{/}\cos\theta = \left(m_0 c + \frac{p_0}{n} - \frac{p_0^{/}}{n} \right)^2 - (m_0 c)^2 \cdot \tag{4.2.25}$$

Carrying out further transformations (4.2.25), we find:

$$p_0 - p_0^{/} = \frac{n}{2m_0 c}\left[\left(p_0^{\ 2} + p_0^{/2}\right)\left(1 - \frac{1}{n^2}\right) + 2p_0 p_0^{/}\left(\frac{1}{n^2} - \cos\theta\right) \right] \cdot \tag{4.2.26}$$

Adding and subtracting units in brackets of the last addend (4.2.26), we shall receive:

$$p_0 - p_0^{/} = \frac{n}{2m_0 c}\left(p_0 - p_0^{/}\right)^2\left(1 - \frac{1}{n^2}\right) + \frac{np_0 p_0^{/}}{m_0 c}(1 - \cos\theta) \cdot \tag{4.2.27}$$

The difference of impulses of falling and scattered photons on electron in vacuum, i.e. at $n = 1$, equals:

$$p_0 - p_0^{/} = \Delta p_{vac} = \frac{p_0 p_0^{/}}{m_0 c}(1 - \cos\theta). \tag{4.2.28}$$

Thus, the formula (4.2.27) can be written down as:

$$\Delta p_0 = \frac{n}{2m_0 c}\Delta p_0^2\left(1 - \frac{1}{n^2}\right) + n\Delta p_{vac}, \tag{4.2.29}$$

where is the difference of impulses falling and scattered photons on electron in substance.

We use dimensionless differences of photon impulses - in substance and - in vacuum. Values and are the frequency change at the scattering of the photon, respectively in substance and in vacuum.

In new designations, the equation (4.2.29) can be copied:

$$\frac{1}{2}n\eta^2\left(1 - \frac{1}{n^2}\right) - \eta + n\eta_{vac} = 0. \tag{4.2.30}$$

Solving a quadratic (4.2.30), we find:

$$\eta = \frac{n}{n^2 - 1}\left[1 - \sqrt{1 - 2\left(n^2 - 1\right)\eta_{vac}}\right]. \tag{4.2.31}$$

The plus before a root is unacceptable. In this case, differences of impulses of photons in substance and in vacuum have a different orientation of change. It will be visible from the further, to use in (4.2.31), expansion of a root as not informatively for the analysis of influence of characteristics of substance on the noncoherent scattering of a photon.

Let's find in an evident form the change of wave length of photons at their noncoherent scattering. We shall transform values:

$$\eta = \frac{\Lambda \Delta \nu_0}{c} \approx -\frac{\Lambda}{c} d\nu_0 = -\frac{\Lambda}{c} d\left(\frac{V_{ph}}{\lambda}\right) = -\frac{\Lambda}{n} d\left(\frac{1}{\lambda}\right) = \frac{\Lambda}{n\lambda^2} d\lambda \approx \frac{\Lambda}{n\lambda^2} \Delta\lambda \qquad (4.2.32)$$

$$\eta_{vac} = \frac{\Lambda \Delta \nu_{vac}}{c} \approx -\frac{\Lambda}{c} d\nu_{vac} = -\frac{\Lambda}{c} d\left(\frac{c}{\lambda_{vac}}\right) = \frac{\Lambda}{\lambda_{vac}^2} d\lambda_{vac} \approx \frac{\Lambda}{\lambda_{vac}^2} \Delta\lambda_{vac}, \qquad (4.2.33)$$

where λ and are wave lengths at the scattering of photons on electrons in substance and in vacuum. Minus signs are connected by the values and are accepted as positive. Besides, it is designated and $\Delta\lambda_{vac} = \lambda_{vac}^{/} - \lambda_{vac} \approx d\lambda_{vac}$, where strokes mean wave lengths of photons after interaction with the electron.

Using a determination of a parameter of refraction as [23], we shall transform (4.2.33) to a form:

$$\eta_{vac} \approx \frac{\Lambda}{n^2 \lambda^2} \Delta\lambda_{vac}. \qquad (4.2.34)$$

Substituting (4.2.32) and (4.2.34) in (4.2.31), we shall find:

$$\Delta\lambda = \frac{n^2}{n^2 - 1} \frac{\lambda^2}{\Lambda} \left[1 - \sqrt{1 - 2\frac{\left(n^2 - 1\right)}{n^2} \frac{\Lambda}{\lambda^2} \Delta\lambda_{vac}} \right]. \qquad (4.2.35)$$

In expansion of a root in the formula (4.2.35) up to the first order, we have $\Delta\lambda = \Delta\lambda_{vac}$; therefore, it is necessary to do the expansion up to the second order $\sqrt{1-x} \approx 1 - \frac{1}{2}x - \frac{1}{8}x^2$. As a result, we shall find:

$$\Delta\lambda \approx \Delta\lambda_{vac} + \frac{1}{2}\left(\frac{n^2 - 1}{n^2}\right) \frac{\Lambda}{\lambda^2} \Delta\lambda_{vac}^2, \qquad (4.2.36)$$

where $\Delta\lambda_{vac} = \Lambda(1 - \cos\theta)$; for example [25].

Thus, in the second order of expansion (4.2.35) the change of wave length to a non coherent scattering depends on the wave length.

It should be noted that in the X-ray area of a radiation spectrum a parameter of refraction $n<1$. Therefore, according to (4.2.36) $\Delta\lambda < \Delta\lambda_{vac}$.

Following [23], let's use a unit decrement of a refraction parameter $\delta = 1 - n$. Then the formula (4.2.36) will be transformed to a form:

$$\Delta\lambda \approx \Delta\lambda_{vac} - \frac{1}{2}\frac{\delta}{\lambda^2}\left(\frac{n+1}{n^2}\right)\Lambda\Delta\lambda_{vac}^2. \tag{4.2.37}$$

The refraction parameter of substances in the X-ray area is close to one [23]; therefore, the formula (4.2.37) can be copied as follows:

$$\frac{\Delta\lambda_{vac} - \Delta\lambda}{\Delta\lambda_{vac}} = \frac{\delta}{\lambda^2}\Lambda\Delta\lambda_{vac}. \tag{4.2.38}$$

The formula (4.2.38) shows that in the X-ray area the influence of the substance on parameters of non coherent scattering is very insignificant and in practical calculations it is possible to use the classical Compton's formula $\Delta\lambda \approx \Delta\lambda_{vac} = \Lambda(1-\cos\theta)$. The opportunity to use Compton's formula in substance is defined mathematically by that expansion (4.2.35); up to the first order the parameters of substance are not present.

Proceeding from the classical theory of a Scattering of Lorentz [23], we have in the X-ray area dependence:

$$\frac{\delta}{\lambda^2} = 2.7 \cdot 10^{11}\frac{\rho Z}{A}\left[m^{-2}\right], \tag{4.2.39}$$

where ρ is the density of the substance in kg/m^3, Z – the charging number, A – the mass number. Using $\Lambda = 2.426 \cdot 10^{-12}\,m$, we find:

$$\tau = \frac{\Delta\lambda_{vac} - \Delta\lambda}{\Delta\lambda_{vac}} = 0.655\frac{\rho Z}{A}\Delta\lambda_{vac}. \tag{4.2.40}$$

The relative change of shift of wave length τ in the substance in comparison with the vacuum at noncoherent scattering in the X-ray area of the electromagnetic radiation spectrum is proportional to the density of the substance, its charging number and in inverse proportion to the nuclear mass of the substance. It also linearly depends on Compton's shift of wave length in vacuum.

4.2.4. Stopping Free Electron After Interaction with A Photon

Let's find conditions at which, after interaction with a photon, the free electron in substance will be stopped; i.e. $p' = 0$.

In this case, the equation (4.2.5) will look like:

$$\frac{pc}{V} + \frac{p_0}{n} = m_0 c + \frac{p_0'}{n}. \tag{4.2.41}$$

From the equation (4.2.41) we find:

$$p = \left(nm_0 c - \Delta p_0\right)\frac{V}{nc}, \tag{4.2.42}$$

where, as before $\Delta p_0 = p_0 - p_0'$.

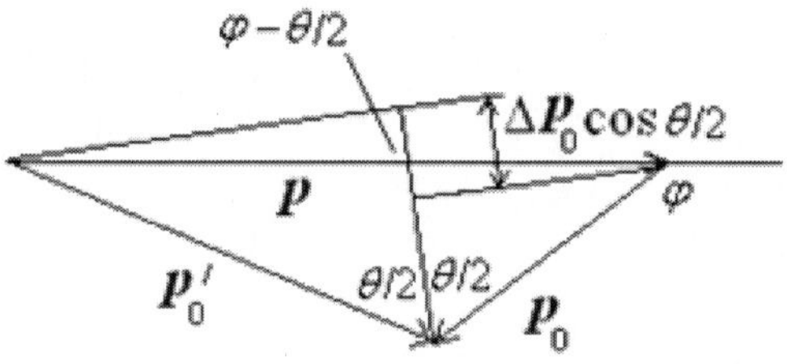

Figure 4.2.4.

Formulas (4.2.4) and (4.2.12) are changed:

$$p_0^{\,2} + p_0'^{\,2} - 2p_0 p_0' \cos\theta = p^2, \tag{4.2.43}$$

$$p^2 + p_0^2 + 2pp_0 \cos\varphi = p_0'^{\,2}. \tag{4.2.44}$$

Using $p_0' = p_0 - \Delta p_0$, the equation (4.2.44) is possible to transform to a form:

$$p^2 + \Delta p_0\left(2p_0 - \Delta p_0\right) + 2pp_0 \cos\varphi = 0. \tag{4.2.45}$$

Solving in common (4.2.43) and (4.2.44) we shall find:

$$p_0 + p\cos\varphi - p_0^{/}\cos\theta = 0, \tag{4.2.46}$$

that is equivalent to:

$$p_0(1-\cos\theta) + p\cos\varphi + \Delta p_0\cos\theta = 0. \tag{4.2.47}$$

Finding from (4.2.47) value and substituting it in (4.2.45), we have:

$$p^2 - \Delta p_0{}^2 - \frac{2p^2\cos^2\varphi}{(1-\cos\theta)} - \Delta p_0\frac{2p\cos\theta\cos\varphi}{(1-\cos\theta)} - \Delta p_0\frac{2p\cos\varphi}{(1-\cos\theta)} - \Delta p_0{}^2\frac{2\cos\theta}{(1-\cos\theta)} = 0 \tag{4.2.48}$$

Transforming (4.2.48), we find:

$$\Delta p_0{}^2 - 2\Delta p_0 p\cos\varphi - p^2\left(1 - \frac{2\cos^2\varphi}{(1-\cos\theta)}\right)\frac{(1-\cos\theta)}{(1+\cos\theta)} = 0 \tag{4.2.49}$$

Solving a quadratic (4.2.49) we shall receive:

$$\frac{\Delta p_0}{p} = \cos\varphi + \sqrt{(\cos\varphi)^2 + \frac{(1-\cos\theta)}{(1+\cos\theta)}\left(1 - \frac{2\cos^2\varphi}{(1-\cos\theta)}\right)}. \tag{4.2.50}$$

As will be shown below, before a root it is necessary to take a plus sign.
After simple transformations, the formula (4.2.50), we shall find:

$$\frac{\Delta p_0}{p} = \frac{\cos\left(\varphi - \frac{\theta}{2}\right)}{\cos\frac{\theta}{2}}. \tag{4.2.51}$$

The formula (4.2.51) shows that at a head-on collision of a photon and a moving electron, i.e. at in case the electron after interaction does stop, a direction of a scattered photon equiprobably in all sides.

There is a plus sign before a root in the formula (4.2.50) connected to geometrical interpretation of the formula (4.2.51), figure 4.2.4.

Let's note that in case the electron after interaction with a photon does not stop, the formula (4.2.51) according to the similar simple geometrical analysis fig. 4.2.2, has a more general view:

$$\left|\Delta p_0\cos\frac{\theta}{2}\right| = \left|p'\cos\left(\gamma - \frac{\theta}{2}\right) - p\cos\left(\varphi - \frac{\theta}{2}\right)\right|. \tag{4.2.52}$$

Using (4.2.42) formula (4.2.51) will be transformed to a form:

$$\Delta p_0 = \frac{\cos\left(\varphi - \frac{\theta}{2}\right)}{\cos\frac{\theta}{2}} \frac{(m_0 V)}{\left[1 + \frac{\cos\left(\varphi - \frac{\theta}{2}\right)}{\cos\frac{\theta}{2}} \frac{V}{nc}\right]} . \qquad (4.2.53)$$

Similarly (4.2.32), we shall transit from the change of impulse of photons to a shift of wave lengths:

$$\frac{\Delta p_0}{m_0 c} = \frac{\Lambda \Delta \nu_0}{c} \approx -\frac{\Lambda}{c} d\nu_0 = \frac{\Lambda}{n\lambda^2} \Delta\lambda . \qquad (4.2.54)$$

Hence, from (4.2.53) we shall find:

$$\Delta\lambda = \frac{n\lambda^2 \Delta p_0}{\Lambda m_0 c} = \frac{\lambda^2}{\Lambda} \frac{\cos\left(\varphi - \frac{\theta}{2}\right)}{\cos\frac{\theta}{2}} \frac{nV/c}{\left[1 + \frac{\cos\left(\varphi - \frac{\theta}{2}\right)}{\cos\frac{\theta}{2}} \frac{V}{nc}\right]} . \qquad (4.2.55)$$

The formula (4.2.55) is a reflection of a special case of noncoherent scattering when the electron on which there is a scattering, after interaction with a photon, stops. The energy of the movement of the electron is transferred to a scattered photon; therefore, its frequency increases, and the wave length decreases in size $\Delta\lambda$.

As the case when the electron stops is examined, the angle $90^0 < \varphi < 270^0$, fig. 4.2.4. The average size of the angle φ takes place in the case of a head-on collision of a photon of X-ray radiation ($n \approx 1$) and electron; therefore, we shall be limited to this case. At $\varphi = 180^0$, we have:

$$\Delta\lambda = -\frac{\lambda^2}{\Lambda} \frac{V/c}{\left(1 - \frac{V}{c}\right)} . \qquad (4.2.56)$$

The minus sign specifies a reduction in wave length after the collision of a photon with a stopping electron.

Let's note that the formula (4.2.56), really, can give an idea to describe the Compton-effect Doppler's shift of wave length [26]. However, dependence of the shift of wave length on angle θ in the formula (4.2.55) differs from those in [26] which, proceeding from the Doppler-effect, is received as $\Delta\lambda \sim (1-\cos\theta)$.

The change of shift of wave lengths due to the movement of the electron at noncoherent scattering, taking into account (4.2.56), is equal to:

$$\delta\lambda \approx \Delta\lambda_{vac} - \frac{\lambda^2}{\Lambda}\frac{V/c}{\left(1-\frac{V}{c}\right)} = \Lambda\left[1-\cos\theta - \left(\frac{\lambda}{\Lambda}\right)^2 \frac{V/c}{\left(1-\frac{V}{c}\right)}\right]. \tag{4.2.57}$$

In figure 4.2.5 the graph of dependence of shift of the noncoherent line of scattered radiation from angle θ [23] is shown. The experimental points received for carbon at the wave length of radiation are also shown. From the analysis of the graph it follows that at the angle θ close to , a rather big reduction of shift is observed. This reduction can be explained only by the movement of the electron since the formula (4.2.37) cannot describe such a big reduction $\Delta\lambda$. A delay of the electron movement at the interaction with a photon results in an increase in energy of the scattered photon, i.e. a reduction value in $\delta\lambda$. The estimation which has been carried out under the formula (4.2.57) shows that the size observable in the experiment at the angle arises at an initial electron velocity $V = 0.6\cdot 10^6\ \frac{m}{s}$.

This speed is close to the Fermi velocity for the electron which can be counted analogue of velocity of thermal electronic gas movement [27].

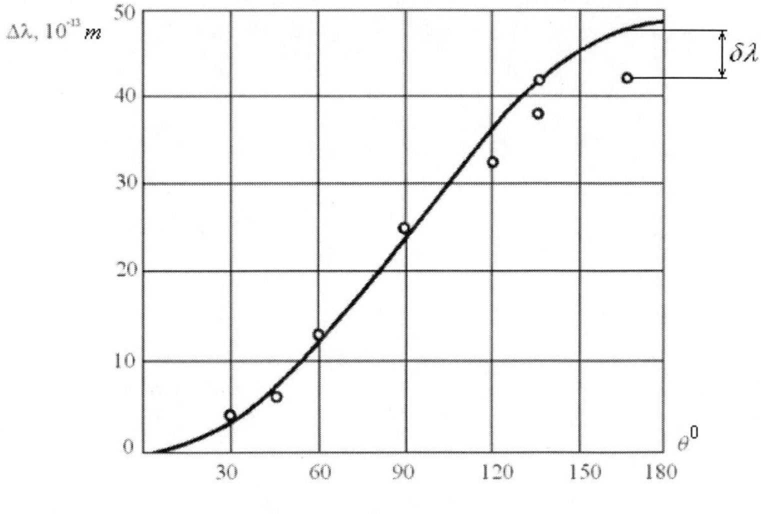

Figure 4.2.5.

Calculation under the formula (4.2.57), obviously, has an estimated character. First, there is the Fermi-Dirac thermal distribution of velocity for electronic gas in substance; therefore, quantums' various energy are necessary for the electron to stop. Secondly, the head-on collision of photons and electron occurs rarely enough. Thirdly, the stopping of the electron after it has interacted with a photon also is a rare enough occurrence. A general change of a photon impulse at noncoherent scattering is defined by the law (4.2.52). Strictly speaking, a stop in the electronic gas two electrons with various directions can only spin. This follows from a Pauli principle of an interdiction. Hence, there should be a statistical distribution of the value which is determined by stochastic ways of formation of its numerical size. However, we can take into account that the head-on collision of a photon and electron is observed when angle φ achieves an average size of in the considered process ($90^o < \varphi < 270^o$). Taking into account also the affinity of the experimental estimation of the electron velocity under the formula (4.2.57) the velocity of Fermi, it is possible to assume that under this formula there we can estimate the rough size of the mathematical expectation of value $\delta\lambda$, in extreme measures, at angles θ close to 180^0.

Thus, at the interaction of a photon with a free electron in substance, various results are possible: the absorption of a photon, coherent and noncoherent scattering, the stopping of the electron.

Thus, the absorption of a photon in the X-ray area probably only in substance with a parameter of refraction is more than one.

The change of a wave length at the noncoherent scattering of X-ray radiation in substance in the second order depends on the wave length, as opposed to the noncoherent scattering in a vacuum. The influence of parameters of substance: density, charging and mass numbers on shift of wave lengths of falling and scattered radiations, very insignificantly.

The observable reduction of shift of wave lengths at a noncoherent scattering can be explained only by the movement of the free electrons interacting with photons. At a head-on collision of a photon and moving electron, in case the electron stops after interaction, a direction of the scattered photon equiprobably in all sides.

Chapter 5

PHYSICAL KINETICS OF INTERACTIONS OF AN ELECTROMAGNETIC FIELD WITH A SUBSTANCE

The theory of the kinetic equations is rather widely applied to the analysis of various physical processes [28]. The interaction of molecules of ideal gas, particles in plasma, phonons in crystals, etc. has been successfully analyzed with the help of this theory. Most strict deductions of Maxwell's distribution of ideal gas molecules by the speeds (or their kinetic energy) is carried out with the help of Boltzmann's kinetic equation [21] in consideration of the system of interacting atoms in a stationary condition. Boltzmann's kinetic equation, first of all, reflects the non-stationary process of the interaction of particles among themselves, the influence of this interaction on the establishment of corresponding distribution of molecules and the influence on the examined system of external factors.

The method of the kinetic equations using a principle of detailed equilibrium is applied also at the research of the transition of the atoms interacting with electromagnetic radiation, from one condition in another [21].

For photons the difference, for example, from molecules of gas, the energy distribution can vary only at their interaction with atoms of substance [21]. Therefore, the character of the transient process to equilibrium radiation should depend on the characteristics of the gas and the distribution of the particles directly interacting with radiation.

In references which investigate the Boltzmann kinetic equation for photons in gas it is possible to allocate two directions. The first direction uses the so-called one-high-speed theory of translation of particles [29, 30, 31]. Most attention in these references is given to the kinetic equation for atoms of substance. The kinetic equation for radiation is written similarly and as a whole has an auxiliary character. The Bose - Einstein transition of radiation to equilibrium distribution is not usually analyzed.

The second direction is connected to A.S. Kompaneets's basic article [32]. Further analysis of the kinetic equation suggested to him is carried out in [33]. There is a main fault of this equation, in our opinion, which is the assumption of Maxwell's function of electron distribution. It is possible to note also that the examined system of photons and electrons' interaction in rarefied gas differs from the similar system in a solid body, where the account of the Pauli principle in the system of electronic gas (besides the function of electrons' distribution in atom) is necessary [4, 34, 35]. It complicates the integrity of the interaction in the kinetic equation.

The purpose of the present chapter is the reception and the analysis of Boltzmann's kinetic equation for system photons - rarefied gas, research of time dynamics of approach of energy photons distribution to equilibrium distribution Bose - Einstein, density of energy of thermal radiation to equilibrium Stefan – Boltzmann law, deviation of non-equilibrium radiation from the fourth degree law of temperature, and also research of other phenomena in radiating gas.

5.1. KINETIC EQUATION OF INTERACTION OF ELECTROMAGNETIC RADIATION WITH RAREFIED GAS

For finding the Boltzmann kinetic equation of system photons - rarefied gas - we shall consider in more detail the transient process of the establishment of equilibrium distribution of quantums of electromagnetic radiation. For this purpose, we shall consider the interaction of some flux of monochromatic quantums of non-equilibrium electromagnetic radiation by energy $\varepsilon = h\nu$, where *h* is Planck's constant, ν - frequency of electromagnetic radiation, with atoms of substance in some volume *V*, figure 5.1.1.

The number of quantums falling on the volume we shall designate n_1; the number of quantums which are taking off from volume *V* in all directions, we shall designate n_3. The system with which the quantums interacts, we shall present in general. We shall assume that it will consist of identical particles - atoms which poorly interact with each other (rarefied gas), and can be in two conditions: particles in the first condition and particles in the second condition.

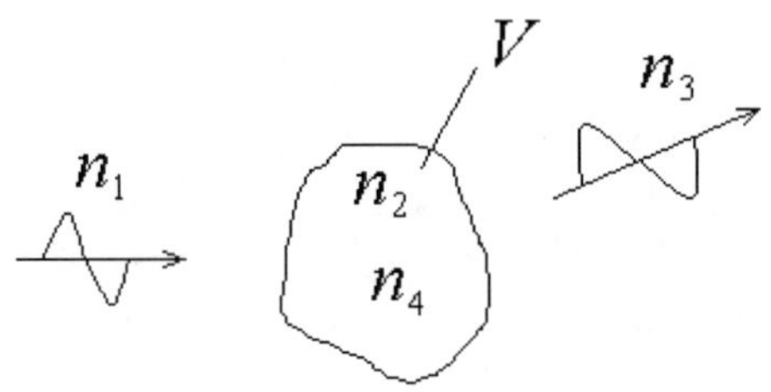

Figure 5.1.1.

A change in the number of photons taken off from the element of volume equal to dn_3 is determined by the following factors: a change in the number of falling photons dn_1, or change in the number of photons due to their interaction with particles in the first condition (absorption of quantums by the non-excited atoms of substance), or change in the number of photons due to their interaction with particles in the second condition (radiations of photons from substance by the excited atoms). Hence:

$$dn_3 = dn_1 + dn_2 + dn_4 . \tag{5.1.1}$$

The formula (5.1.1) is a consequence of the theorem for the probability of the sum of incompatible events. The probability (in this case absolute frequency) of the complex event is equal to the sum of probabilities (absolute frequencies) of incompatible events which can result from the given complex event. We shall also formally attach the radiation of quantum from the atom to the event of interaction of quantum and atom. It is accepted that with one quantum there can be only one of the above-listed events.

We shall present a change in the number of quantums flying in volume as:

$$dn_1 = \frac{2}{h^3} f_1 dp_1 dV , \tag{5.1.2}$$

where is the product of changes in the components of flying quantums' impulse, and - element of phase volume in which there are flying quantums.

The value is the function of distribution of flying quantums, - module of the radius-vector of the flying quantum, - element of the phase volume, occupied by two particles; atom and quantum [36], factor 2 reflects two possible polarizations of photon in volume h^3.

Let's designate the change of the atom impulse due to its excitation by quantum. Obviously, it is under the law of impulse conservation $dp_2 = dp_1$, as it is possible to assume that the interacting quantum and atom are in one point $d\gamma_1 = d\gamma_2$, where is the element of the phase volume in which there are atoms interacting with quantums. From the equation (5.1.1) it is possible to obtain:

$$\frac{dn_3}{d\gamma_1} - \frac{dn_1}{d\gamma_1} = \frac{dn_2}{d\gamma_2} + \frac{dn_4}{d\gamma_2} = \frac{2}{h^3}\theta , \tag{5.1.3}$$

where is some function of coordinates, impulses and time, and index $i = 1, 2$. The factor is used for simplification of transformations.

Using (5.1.2) and (5.1.3), it is possible to write down:

$$dn_3 = \frac{2}{h^3}(\theta + f_1) d\gamma_1 \tag{5.1.4}$$

Having designated:

$$dn_4 = \frac{2}{h^3} f_2 dp_2 dV , \tag{5.1.5}$$

where is the function of distribution of the excited atoms, factor 2 - two possible orientations of the electron spinning in volume h^3; from (5.1.3) and (5.1.5) we shall obtain:

$$dn_2 = \frac{2}{h^3}(\theta - f_2)d\gamma_2. \tag{5.1.6}$$

The number of interactions of the quantums which have flown in volume dV, with non-excited atoms of substance being equal to:

$$dn_1 dn_2 = \frac{4}{h^6} f_1(\theta - f_2)dp_1 dV d\gamma_2. \tag{5.1.7}$$

This number of quantums leaves phase volume at interaction.

On the other hand, in this phase volume the certain number of the taken off quantums gets:

$$dn_3^/ dn_4^/ = \frac{4}{h^6} f_2^/\left(\theta + f_1^/\right)dp_2 dV d\gamma_1. \tag{5.1.8}$$

Strokes at functions of distribution mean that a kind of function of distribution, both photons, and the excited atoms can change during the time before and after the interaction of quantums with atoms.

The rest in the phase volume quantums can be found under the formula:

$$d^2 n = \frac{4}{h^6}\left[f_1(\theta - f_2) - f_2^/\left(\theta + f_1^/\right)\right]\delta\left(h\nu^* - \varepsilon_i^*\right)d\gamma^2. \tag{5.1.9}$$

In the formula (5.1.9) is delta - function which takes into account that only at the energy of quantum equal to the energy of electronic transition is there interaction of quantums and atoms of substance. Asterisks at values mean that the comparison goes on dimensionless energy, which scale is identical. A similar approach, for example, is carried out in [28].

Let's copy the formula (5.1.9) as:

$$df = d\left(\frac{h^3}{2}\frac{dn}{d\gamma}\right) = \frac{2}{h^3}\left[f_1(\theta - f_2) - f_2^/\left(\theta + f_1^/\right)\right]\delta\left(h\nu^* - \varepsilon_i^*\right)d\gamma, \tag{5.1.10}$$

where is the function of distribution of quantums changing in time, before interaction had function of distribution f_1, and after interaction $f_1^/$. Accordingly, for this time the function of distribution of the excited atoms has changed also from up to $f_2^/$. At transition it is in a stationary condition $f \to f_1$.

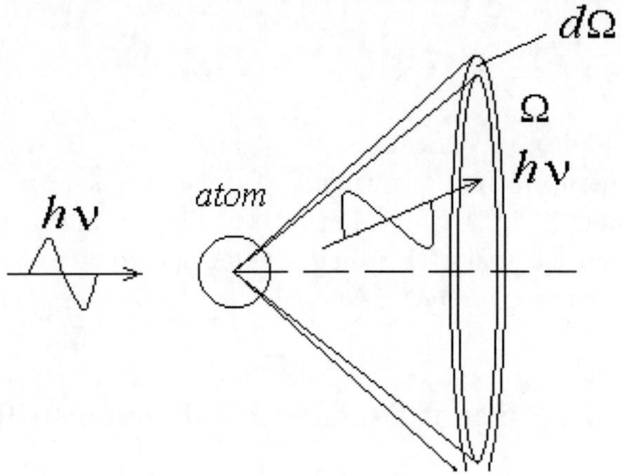

Figure 5.1.2.

We use the effective section of interaction σ of quantum with atom. Then, there is a change of the effective section of interaction depending on a solid angle of quantum scattering Ω, figure 5.1.2; it is possible to write this down as:

$$d\sigma = \sigma d\Omega. \tag{5.1.11}$$

Hence, the element of volume of interaction becomes:

$$dV = cd\sigma = c\sigma d\Omega. \tag{5.1.12}$$

In the formula (5.1.12) the unit time interval is considered; therefore, instead of the length of volume element the speed of quantum c used. Besides, we have neglected the speed of movement of the atom, in comparison with the speed of the quantum of electromagnetic radiation interacting with it.

Taking into account that $p = \frac{h\nu}{c}$, we shall find the element of phase volume:

$$d\gamma = dpdV = \frac{h^3}{c^3} d\nu^3 \sigma c d\Omega = \frac{h^3}{c^2} \sigma d\nu^3 d\Omega. \tag{5.1.13}$$

Integrating the formula (5.1.10) on all frequencies and a solid angle, with the account (5.1.13), we obtain:

$$\Delta f = \frac{2}{c^2} \iint_{\nu,\Omega} \sigma \left[f_1 (\theta - f_2) - f_2^/ \left(\theta + f_1^/\right)\right] \delta\left(h\nu^* - \varepsilon_i^*\right) d\nu^3 d\Omega \cdot \tag{5.1.14}$$

We examine the change of function of quantums' distribution in a unit of time; therefore, in the left part (5.1.10) will be the speed of change of function of quantums' distribution [28]. In the right part (5.1.10) between functions of distribution without strokes and with strokes the time interval also is an equal unit. Hence:

$$\frac{df}{dt} = \frac{2}{c^2} \iint_{\nu,\Omega} \sigma\left[f_1(\theta - f_2) - f_2'\left(\theta + f_1'\right)\right]\delta\left(h\nu^* - \varepsilon_i^*\right)d\nu^3 d\Omega, \tag{5.1.15}$$

where is integral to the interaction of quantum of electromagnetic radiation with atom of substance. The equation (5.1.15) is the kinetic equation of interaction of quantums of electromagnetic radiation and atoms. The solution to the integro-differential equation (5.1.15) is a rather difficult mathematical problem. We shall consider two special cases.

5.1.1. A Condition of Balance at Interaction of Radiation with Gas Atoms

We shall investigate the most simple case – the interaction of quantums of equilibrium electromagnetic radiation and equilibrium system of atoms. In this case, the distribution of quantums and atoms of substance does not vary; i.e. $\frac{df}{dt} = 0$. From this condition, according to (5.1.15), it is possible to conclude, that:

$$f_1(\theta - f_2) - f_2(\theta + f_1) = 0. \tag{5.1.16}$$

The functional equation with two unknown functions which has an infinite set of solutions is received. Strokes in the equation (5.1.16) are omitted, since functions of distribution and in this case in time do not vary.

The second equation to (5.1.15), and, hence, in balance and to (5.1.16), the kinetic equation for electrons (atoms) should be additional. This equation, apparently, should be analogous to the kinetic equation for the interaction of electrons and phonons in metal [4, 34] since phonons and photons are bosons.

We use the known fact that in the case of balance, the function of distribution in the equation (5.1.16) should satisfy the Bose - Einstein distribution for photons, radiated by the absolutely black body [37]:

$$f_1 = \frac{1}{e^X - 1}, \tag{5.1.17}$$

where $X = \frac{\varepsilon_i - \mu}{kT}$, - energy of quantum in phase volume $d\gamma_i$, μ - the chemical potential attributed to one quantum, - the Boltzmann constant, T - absolute temperature of examined quantum system.

Substitution (5.1.17) in (5.1.16) allows for the reception of distribution of the excited atoms:

$$f_2 = \frac{1}{e^X + (2/\theta - 1)}. \tag{5.1.18}$$

At the distribution (5.1.18) will be transformed to the Fermi - Dirac distribution [37]. In this case, the distribution of the excited atoms completely coincides with the distribution of electrons in these atoms. It is natural, since excitation of atom is defined by the transition of electrons on a higher energy level. The value is $X = \frac{\varepsilon_i - \mu}{kT}$. But as opposed to (5.1.17) energy of the excited atom in the phase volume $d\gamma_i$, corresponding to energy of excitation of the electron under action of radiated quantum by energy ε_i, and μ in this case is the energy of the Fermi level.

From (5.1.18) it is also possible to conclude that function θ to aspire to the unit at transition of the system of interaction photon - atom in a stationary condition, i.e. in a case where the function of distribution *f* does not depend on time.

In the carried out analysis it is impossible to believe a deduction of quantum distributions Bose – Einstein and Fermi - Dirac. The functional equation (5.1.16) with two unknown functions of distribution also has an infinite set of solutions; i.e. the equation (5.1.16) indefinitely satisfies plenty of functions of distribution. It is connected to the rather wide spectrum of properties of the examined system of interaction of particles. But the equation (5.1.16) should not fail to satisfy the equilibriums' quantum distributions of photons and electrons. The definition of a function of distribution for quantums (5.1.17) makes the functions of electrons' distribution (5.1.18) more certain. In an uncertain value there is function θ concerning which (5.1.15), apparently, it is impossible to make any conclusions from the analysis of the kinetic equation. But at the same time the definition of function completely determines a kind of argument in function which should be identical to a similar argument in function f_1.

At the distribution of atoms of substance in a condition of balance (5.1.18) has become identical to Maxwell's distribution. This means that quantums of electromagnetic radiation do not cause the excitation of atoms and do not influence their distribution. Distribution of atoms remains constant and corresponds to the distribution of ideal gas.

Thus, it is possible to make the conclusion that function θ reflects the character of interaction of quantums with atoms of substance; in particular, at there is an excitation of atoms; at it is absent.

5.1.2. Transition of Radiation to the Equilibrium Condition

As the second example we shall consider the time asymptote of energy distribution of the photons aspiring to equilibrium radiation of the black body at interaction with the substance. The kinetic equation (5.1.15) can be copied using the expression for a full derivative in the left part:

$$\frac{\partial f}{\partial t}+c\frac{\partial f}{\partial r}+F_e\frac{\partial f}{\partial p}=\frac{2}{c^2}\iint\limits_{\nu,\Omega}\sigma\left[f_1(\theta-f_2)-f_2^{/}\left(\theta+f_1^{/}\right)\right]\delta\left(h\nu^*-\varepsilon_i^*\right)d\nu^3 d\Omega, \quad (5.1.19)$$

where is the external force acting on a photon, for example, from the external gravitational field, c is a velocity of quantum [38].

Let's assume that the function of the excited atoms' distribution is constant in time and is Fermi – Dirac's equilibrium function, $f_2 = f_2^{/}$. The function of photons' distribution we shall present as:

$$f_1^{/} = f_1 + \partial f_1, \quad (5.1.20)$$

where is a small variation of function for small time ∂t. Besides, we believe that with the external force acting on quantums there is no ; also, a variation of the function of quantums' distribution on space is absent $\frac{\partial f}{\partial r}=0$. Substituting (5.1.20) in (5.1.19), we shall obtain:

$$\frac{\partial f_1}{\partial t}=\frac{2}{c^2}\iint\limits_{\nu,\Omega}\sigma\left[f_1(\theta-f_2)-f_2(\theta+f_1)-f_2\partial f_1\right]d\nu^3 d\Omega. \quad (5.1.21)$$

We assume also that the distinction in electronic levels energy is insignificant, that is correct for external electrons of atom. Therefore, the electrons' energy spectrum we believe in the given approximation continuous and delta - function in (5.1.21) is not taken into account. For variable size of function of quantums' distribution in the left part (5.1.21) for definiteness we use the index 1. The received kinetic equation (5.1.21) reflects the time process of variation of function of quantums distribution at a small deviation of system from the equilibrium condition. The first difference in the integral of interaction in the right part (5.1.21) is identically equal to zero, see (5.1.16), since in this difference only equilibrium functions of distribution are used.

Using for an independent variable - time equality of differential and an increment $\partial t = \Delta t$, we obtain:

$$\frac{\partial f_1}{\partial t}=\frac{2}{c^2}\iint\limits_{\nu,\Omega}\sigma\left[-f_2\partial f_1\right]d\nu^3 d\Omega=-\frac{2}{c^2}\iint\limits_{\nu,\Omega}\sigma f_2\frac{\partial f_1}{\partial t}\Delta t d\nu^3 d\Omega. \quad (5.1.22)$$

Let's designate $\frac{\partial f_1}{\partial t}=F$. Besides, we shall assume that the scattering of photons after interaction with the atom is uniform in all directions; i.e. the section of interaction σ does not depend on the scattering angle. Then, integrating on solid angle Ω, we have:

$$F = -\frac{8\pi}{c^2}\sigma f_2 \Delta t \int_0^{\nu} F d\nu^3. \tag{5.1.23}$$

The homogeneous linear integrated equation with a variable top limit which can be solved concerning function F is received. Finding a third derivative on frequency ν from the left and right parts of the equation (5.1.23), we obtain:

$$\frac{\partial^3 F}{\partial \nu^3} = -\frac{8\pi}{c^2}\sigma f_2 F \Delta t. \tag{5.1.24}$$

The solution of the equation (5.1.24) we shall write down as:

$$F = F_0 \exp\left(-\alpha\sqrt[3]{\Delta t}\right), \tag{5.1.25}$$

where is the constant of integration, value $\alpha = 2\nu\left(\frac{\pi}{c^2}\sigma f_2\right)^{\frac{1}{3}}$.

Solutions (5.1.25) growing on time are rejected, since they reflect the process of system distance from an equilibrium condition. The increase in the non-equilibrium character of the system can occur only with the existence of external influences on it. For example, if gas is compressed in some vessel or on it, the mechanical wave is spread.

Accepting $F = \frac{\Delta f_1}{\Delta t}$, we shall find:

$$\Delta f_1 = F_0 \Delta t \exp\left(-\alpha\sqrt[3]{\Delta t}\right). \tag{5.1.26}$$

The value is the time from the initial moment t_0. If we accept $t_0 = 0$, and is the Bose – Einstein distribution (redesignation of values is made to exclude strokes) also $\Delta f_1 = f_1 - f_{10}$, achievement by distribution of balance condition can be calculated under the formula:

$$f_1 = f_{10} + F_0 t \exp\left(-\alpha\sqrt[3]{t}\right), \tag{5.1.27}$$

where $F_0 = \left(\frac{\partial f_1}{\partial t}\right)_{t=0}$.

Size because the probability of the distribution equilibrium of quantums' energy is greater than non-equilibrium; hence, $f_1 < f_{10}$. Besides, $\Delta f_1 = f_1 - f_{10} < 0$.

The formula (5.1.27) shows how non-equilibrium distribution of photons' energy approaches equilibrium distribution of radiation of Bose – Einstein absolutely black body. It actually gives as a first approximation the time non-stationary amendment to this distribution. At time *t* aspiring to infinity from (5.1.27), naturally, follows $f_1 = f_{10}$.

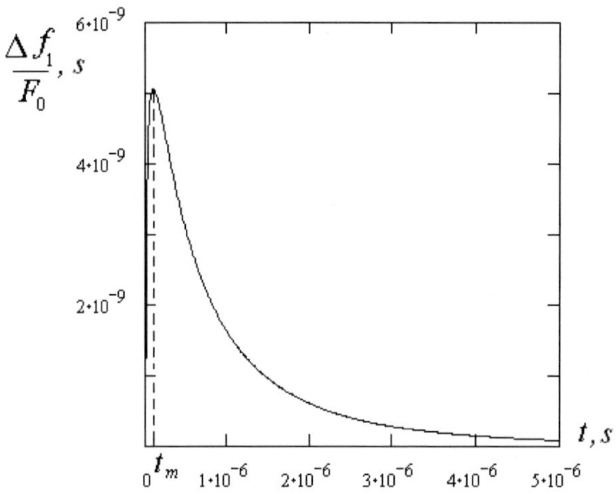

Figure 5.1.3.

On figure 5.1.3 time dependence as function of time *t* is shown. The ratio $\frac{\Delta f_1}{F_0} > 0$, because of the negative sizes in the numerator and denominator. In plotting the curve it is taken into account that function achieves the maximum at the moment of time $t_m = \frac{27}{\alpha^3}$. Apparently, the descending part of the curve has physical sense only. It is connected by that on time the value should decrease up to zero.

Let's estimate size of time t_m, from which the law (5.1.27) for a seen range of waves lengths is correct; for example, ($\lambda = 500\, nm$). We shall assume also $f_2 = 0{,}5$. The effective section of interaction σ can be less than the size of an atom (due to tunnel effect) or greater than an atom (antistokes interaction). For estimation, we shall assume that the effective section of interaction is approximately equal to the sizes of an atom of hydrogen. Hence, $\sigma = 0.878 \cdot 10^{-20}\, m^2$. As a result, it is possible to find parameter $\alpha = 2\nu\left(\frac{\pi}{c^2}\sigma f_2\right)^{\frac{1}{3}} = 642\, s^{-\frac{1}{3}}$; thus, time $t_m = 10^{-7}\, s$. For these parameters the curve on fig. 5.1.3 is plotted. To the law of Stefan - Boltzmann, the finding of the non-stationary amendment is of interest for the volumetric density of energy of thermal radiation *u*.

Using the known formula for volumetric density of energy of thermal radiation [21], after simple transformations (for photons [37]) with use (5.1.27), we shall find:

$$u = \int_0^\infty \frac{8\pi h}{c^3} f_1 \nu^3 d\nu = \sigma_c T^4 + \frac{\beta(T)}{\sqrt[3]{t}}, \tag{5.1.28}$$

where is the constant of Stefan - Boltzmann. The value:

$$\beta(T) = \frac{3\pi h F_0}{c^3 \left(\pi \sigma f_2 / c^2\right)^{4/3}} = \frac{3hF_0}{\left(\pi c \sigma^4\right)^{1/3}} \left[\exp\left(\frac{\varepsilon_i - \mu}{kT}\right) + 1\right]^{4/3}$$

depends on the temperature since function depends on the temperature (5.1.18).

Size due to the negative value F_0. It reflects the known fact that the volumetric density of energy of equilibrium radiation of the black body is the greatest possible.

In time, at the transition of the system in equilibrium condition, the non-stationary amendment in (5.1.28) to aspire to zero, and the volumetric density of energy of radiation to Stefan-Boltzmann. With the growth of temperature the contribution of the non-stationary amendment is also reduced, and the value aspires to the constant $\frac{3hF_0}{\left(\pi c \sigma^4\right)^{1/3}}$.

5.2. Thermal Electromagnetic Radiation of Rarefied Gas

The theory of thermal radiation represents one of the most advanced physical theories. Actually, this theory has begun the development of quantum mechanics. The matter is that even before the creation of quantum mechanics, the formula for spectral density of energy of equilibrium thermal radiation, the so-called Rayleigh – Jeans law has been found. In spite of the fact that the deduction of this law is completely transparent, it was in absolute contradiction with the experiment, showing that the density of energy of electromagnetic radiation monotonously grows with increasing frequency. This has induced M. Planck to undertake research which led him to Planck's well-known formula for radiation of the absolutely black body and the further development of quantum statistics.

All events of this history have been repeatedly discussed in scientific and scientific-popular literature.

We shall consider some questions of thermal radiation to which researchers usually pay insufficient attention.

5.2.1. Processes in the Gas, Resulting to Thermal Electromagnetic Radiation

It is interesting to consider in more detail the processes in the gas resulting in thermal electromagnetic radiation.

In figure 5.2.1 some constant volume of gas *V*, which is heated up due to the submission of heat Q , is shown. It results in the radiation by gas of quantums of electromagnetic energy $h\nu$.

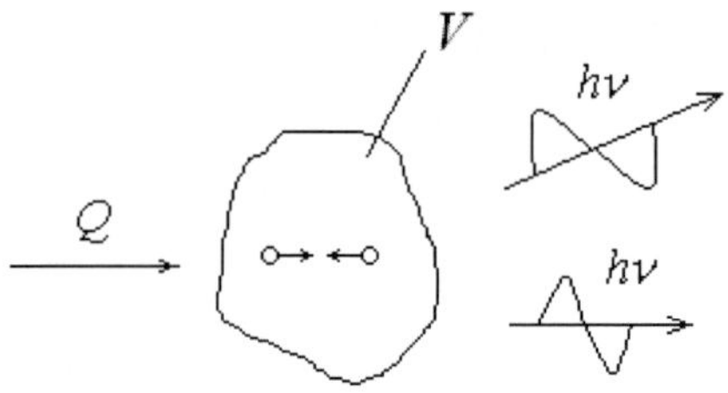

Figure 5.2.1.

Let's carry out the kinetic analysis of processes in the rarefied gas, resulting in electromagnetic radiation.

We shall assume the gas in which the closed thermodynamic system is taking place is in a state of equilibrium. I.e. however many thermal energy Q gives in the system, as much electromagnetic energy is radiated from it. The external interaction of molecules' with electrons in such a gas occurs only at their collision.

Let's consider the probability of transition of electron in a molecule of gas from the top energy level on bottom. This probability is similar to a method of Einstein [21]; we shall divide into two parts, on the spontaneous and compelled probabilities. We shall consider that the compelled transition of electrons occurs only due to the collision of molecules, figure 5.2.2.

$$p_{21} = p_{21}^{s} + p_{21}^{c}. \qquad (5.2.1)$$

Let's note that as at the spontaneous transition of electrons on the bottom energy level, and as compelled there is radiation of electromagnetic energy quantums $h\nu$.

Using a principle of detailed balance and Boltzmann distribution, we have:

$$\frac{p_{12}}{p_{21}} = \exp\left(-\frac{\Delta W}{kT}\right), \qquad (5.2.2)$$

where is the probability of a transition of electrons on the top energy level at impact of molecules, k – Boltzmann constant, T - absolute temperature of gas. The value takes into account the change of the energy condition of molecules at their collision, i.e. redistribution of kinetic energy of molecules and transitions of electrons in molecules from a level on a level.

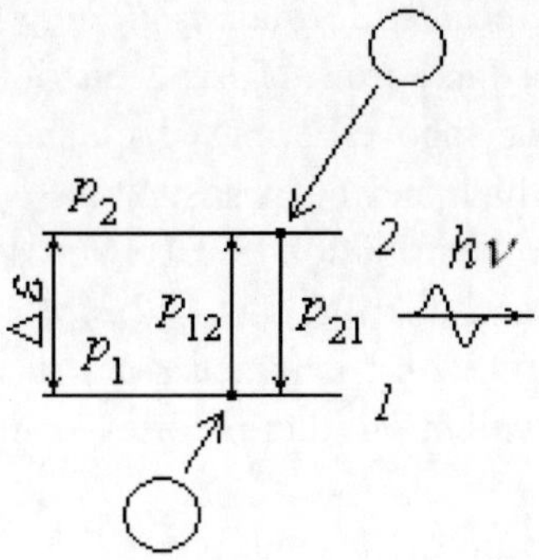

Figure 5.2.2.

Substituting (5.2.1) in (5.2.2), and taking into account that the probability of the compelled transition at the collision of molecules with the bottom energy level on top is equal to the probability of the compelled transition at collision of molecules with the top level on the bottom $p_{12} = p_{21}^{c}$, we find:

$$p_{21}^{c} = \frac{p_{21}^{s}}{\exp\left(\frac{\Delta W}{kT}\right) - 1}. \tag{5.2.3}$$

A change of the energy condition of the molecule generally can take place or due to the transfer of kinetic energy from one molecule to another at their collision, or due to the supply of energy from the outside, or due to the electromagnetic radiation of gas, or all of these cases simultaneously taking place.

On the other hand, at the impact of molecules, in them there is a distribution of external electrons on their energy. This distribution, as well as Maxwell's distribution, has a thermal character. Since it results from collisions of molecules the internal electrons of molecules do not participate in the occurrence of the given distribution.

The probability for the electron to appear at a level 1, figure 5.2.2, we shall accept equals p_1, and at a level 2 equals p_2. Taking into account the condition of normalization and Boltzmann's distribution $\frac{p_2}{p_1} = \exp\left(-\frac{\Delta\varepsilon}{kT}\right)$, we obtain:

$$p_2 = \frac{1}{1 + \exp\left(\frac{\Delta\varepsilon}{kT}\right)}. \tag{5.2.4}$$

The value takes into account only the difference of the electrons' energy at energy levels 1 and 2.

In the form of a record the received formula is close to Fermi – Dirac's [37] distribution, but not identical to it. As opposed to Fermi – Dirac's distribution, in the formula (5.2.4) value is not in excess of electron energy above Fermi's level, and represents a difference of energy levels of external electrons to which they can transit at the collision of molecules. Besides, we shall note that at the deduction of distribution (5.2.4) Pauli's principle was not used.

The Maxwell distribution and distributions (5.2.3), (5.2.4) contradict each other; since there were distributions (5.2.3), (5.2.4), it is necessary that impacts of molecules were not elastic, and to Maxwell's distribution there corresponds absolutely elastic impacts of molecules.

Besides the distinction of forms of distributions (5.2.3) and (5.3.4), concerning as a whole the same system - pays attention to rarefied gas. Hence, the distribution of molecules on energy or to velocities in the examined gas system cannot be described by the uniform formula.

5.2.2. The Quantum Kinetic Equation of Radiating Rarefied Gas

To find out how molecules of gas should be distributed on energy in view of quantum processes, we shall consider the kinetic Boltzmann equation as [21]:

$$\frac{df}{dt} = \iint_{\boldsymbol{p}_1, \Omega_1} v_{rel}\sigma\left(f_2 f_3 - f f_1\right) d\boldsymbol{p}_1 d\Omega_1 \,, \tag{5.2.5}$$

where f and are the density of probabilities of molecules' distribution in the velocities (or energy) for two molecules to have velocities before collision and v_1, and f_2, - density of probabilities for the same molecules to have velocities after collision and v_3. Besides sizes - relative speed of impacting molecules, σ - effective section of interaction of particles, - impulse of examined impacting molecule, - a solid angle in which particles scattered, in this case molecule *a*, after collision, figure 5.2.3.

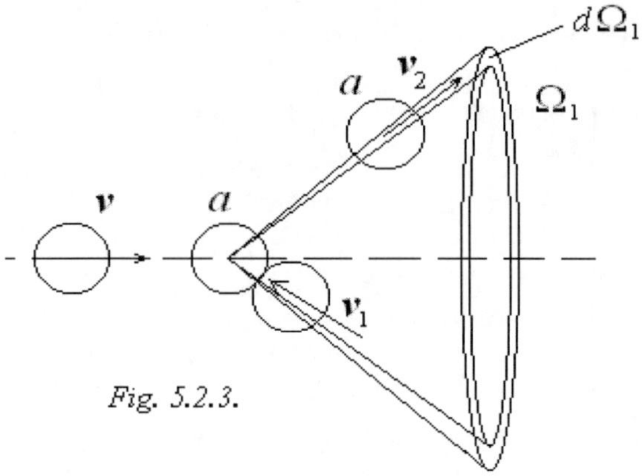

Fig. 5.2.3.

Figure 5.2.3.

The main lack of the Boltzmann equation (5.2.5) is the impossibility of using it to describe the electromagnetic thermal radiation of gas. It is connected to the assumption of absolutely elastic impacts of gas molecules. We shall notice also that this equation, found before the creation of quantum mechanics, has no quantum character.

With the purpose of the account of the non-elastic process of collisions of molecules, finally, resulting in radiation of quantums of electromagnetic energy, and also the translation of Boltzmann's equation in quantum area, we modify the equation (5.2.5). The energy of a molecule represents the sum of kinetic energy of its movement and potential energy of excitation due to the transition of electrons on higher energies' levels [39].

Let's allocate in all the volume of gas the relative share of molecules θ (concerning corresponding interval of energy) which have energies W and W_1. As a result of their collisions there are molecules with energies and W_3, which share too θ. The number of collisions of molecules with energies W and is proportional to the product of probabilities ff_1. But after the first collision the share of molecules with these energies will decrease due to the occurrence of molecules with energies and W_3.

Therefore, the number of collisions will be proportional to the reduction of a share of molecules with energies W and due to the increase in a share of molecules with energies and W_3, i.e. proportional to value $(\theta - f_2 - f_3)$. Hence, Boltzmann's kinetic equation (5.2.5) for radiating gas gets a kind:

$$\frac{df}{dt} = \iint_{p_1,\Omega_1} v_{rel}\sigma\left[f_2 f_3(\theta + f + f_1) - ff_1(\theta - f_2 - f_3)\right]dp_1 d\Omega_1 \cdot \tag{5.2.6}$$

The value shows that the share of molecules with energies and grows due to the molecules which, before collision, had energies W and W_1. In the equation (5.2.6), as opposed to (5.2.5), an integral part of the collisions is written down not only on the basis of balance of molecules before and after collision, but also the dynamics of change of molecules' shares is taken into account during the collisions, resulting in thermal electromagnetic radiation.

Let's find the stationary distributions of the molecules f_i, submitting to the equation (5.2.6). Stationary distributions of molecules arise in a case $\frac{df}{dt} = 0$. Hence, from (5.2.6) we find:

$$f_2 f_3(\theta + f + f_1) = ff_1(\theta - f_2 - f_3), \tag{5.2.7}$$

To the functional equation (5.2.7) satisfies the functions of distribution of type:

$$f = \frac{\theta}{\exp\left(\frac{W}{kT}\right) - 1}, \; f_1 = \frac{\theta}{\exp\left(\frac{W_1}{kT}\right) - 1}, \tag{5.2.8}$$

$$f_2 = \frac{\theta}{1 + \exp\left(\frac{W_2}{kT}\right)}, \; f_3 = \frac{\theta}{1 + \exp\left(\frac{W_3}{kT}\right)}. \tag{5.2.9}$$

Simply to show that the equation (5.2.7) will identically be transformed to a kind:

$$\frac{f_2 f_3}{(\theta - f_2)(\theta - f_3)} = \frac{f f_1}{(\theta + f)(\theta + f_1)}, \tag{5.2.10}$$

which decision, obviously, are functions (5.2.8) and (5.2.9).

Thus, the law of energy conservation is carried out at the non-elastic collision of molecules:

$$W + W_1 = W_2 + W_3 . \tag{5.2.11}$$

Substituting $W = \frac{mv^2}{2} + \Delta\varepsilon$, we find:

$$\frac{mv^2}{2} + \Delta\varepsilon + \frac{mv_1^2}{2} + \Delta\varepsilon_1 = \frac{mv_2^2}{2} + \Delta\varepsilon_2 + \frac{mv_3^2}{2} + \Delta\varepsilon_3 \cdot \tag{5.2.12}$$

In the formula (5.2.12) values characterize the change of molecules' energy to the transition of electrons from level on level. Thus, this transition can be both compelled at impact, and spontaneous.

The kind of functions of molecules' distribution at collision changes from the form (5.2.8) to the form (5.2.9) due to the excitation of molecules. But after radiation of electromagnetic energy by the molecules their function of distribution comes back to the form (5.2.8), figure 5.2.4.

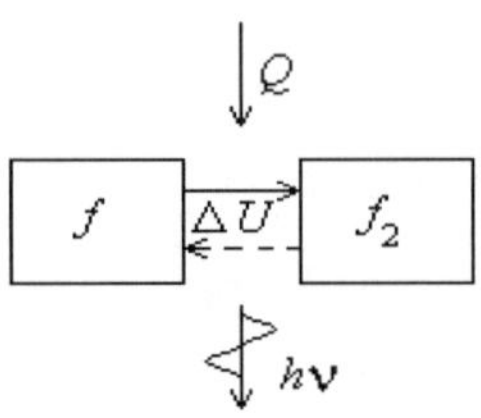

Figure 5.2.4.

In the figure it is marked that it is heat Q, submitted to isovolumetric heated up gas goes on to change the internal energy of gas that results in a change of the law of distribution of gas molecules at their excitation and to radiation of electromagnetic energy. Thus, non-excited molecules have function of distribution (5.2.8), and excited molecules have a function of distribution (5.2.9). In gas, simultaneously, there are molecules distributed under forms (5.2.8) and (5.2.9).

The kinetic equation (5.2.6) has an irreversible character on time that reflects the second law of thermodynamics. Really, collisions of molecules with energies and will not lead to molecules with energies W or W_1. Mathematically, it is connected by the left part of the difference in the integrity of collisions (5.2.6) is not symmetrical to the right part, as opposed to the Boltzmann equation (5.2.5). It is connected by the fact that a fraction of the molecules' energy after their collision is irreversibly lost due to radiation.

Thus, the distribution of molecules in the velocities in the case of their non-elastic impacts cannot be described as one function of distribution.

The distribution of molecules on energies in the given problem where it is examined is not isolated, but closed isovolumetric molecules system submits to laws (5.2.8) and (5.2.9). Thus, the given system of molecules can be both heated up from an external source and emit electromagnetic thermal radiation. But the examined thermodynamic system of molecules is not open since a change of mass is excluded.

Concerning value θ in the equation (5.2.6), it is necessary to make the following explanations. In the kinetic equation (5.2.6) only the interaction of molecules with energies W and is examined. Before the first collision of molecules, information about the resulting system of molecules with energies and W_3, for example, not their relative shares in gas system. But at the first collision of molecules with energies W and , there should already be information on the relative shares of molecules with energies and W_3. Therefore, the factor is used. Actually, this factor before collision of molecules with energies W also takes into account the information on the condition of molecules' system with energies and W_3. From this point of view it is a quantum explanation of a factor $(\theta - f_2 - f_3)$.

For the account of Pauli principle it is necessary to use the multiplier $(1 - f_2 - f_3)$. The point is that at the collision of molecules, there is the interaction of their external electronic shells and Pauli's principle starts to act. Due to this multiplier the number of molecules transit in the filled conditions for which $f_2 + f_3 = 1$, it is equal to zero. I.e. collisions can result in molecules only in as yet unoccupied conditions [28, 34]. In other words, at the fulfilment of a condition , there cannot be a molecule with energies or W_3. This position should be reflected in the equation (5.2.6).

Thus, quantum consideration shows that size $\theta = 1$. It also means that in rarefied gas, by consideration of collisions of molecules with energies W and W_1, there is no necessity to consider other molecules with other energies and a relative share of molecules with these energies including the first collision, it is necessary to accept it will equal one. From the quantum point of view the information on the condition of molecules with energies also is determined by Pauli's principle. Therefore, classical and quantum explanations of the presence of the factor actually containing the information on the condition of system of interacting molecules after collision, it is equally correct.

5.2.3. Distributions of Gas Molecules in the Velocities in Radiating Gas

Taking into account the duality of the function of molecules' distribution on energies, we shall estimate functions of distribution of molecules in the velocities. First, we shall consider a case when the form of distribution corresponds (5.2.9), i.e. molecules are in the excited condition. The probability of finding a molecule energy *W* in phase volume can be received under the formula:

$$dP = f\frac{d\gamma}{\gamma_1} = f\frac{dXdYdZdp_X dp_Y dp_Z}{\gamma_1} = f\frac{dV4\pi p^2 dp}{\gamma_1}, \tag{5.2.13}$$

where is the element of the phase volume, - the volume, falling one molecule in gas, - product of components of molecule impulse, p - the module of molecule impulse, - phase volume falling one molecule. The function of distribution according to (5.2.9) is $f = \dfrac{1}{1+\exp\left(\dfrac{mv^2/2+\Delta\varepsilon}{kT}\right)}$. In the formula (5.2.13), the integration is executed of impulses on angles of spherical layer by thickness that is equivalent to replacement [37].

Transit in (5.2.13) from the impulse to velocity of molecule v, we shall find:

$$dP = f\frac{V4\pi m^3 v^2 dv}{\gamma_1} = \frac{V4\pi m^3}{\gamma_1}\frac{v^2 dv}{1+\exp\left(\dfrac{mv^2/2+\Delta\varepsilon}{kT}\right)} \tag{5.2.14}$$

In the formula (5.2.14) integration of the volume of gas V is carried out.

Taking into account , where is the number of molecules with a speed from v up to $v+\Delta v$, and n the general number of molecules in volume V, from (5.2.14) we shall obtain:

$$dn = F\frac{v^2 dv}{1+\exp\left(\dfrac{mv^2/2+\Delta\varepsilon}{kT}\right)}, \tag{5.2.15}$$

where the constant can be found from the condition of normalization:

$$n = F\int_0^{\infty}\frac{v^2dv}{1+\exp\left(\frac{mv^2/2+\Delta\varepsilon}{kT}\right)}. \tag{5.2.16}$$

The integral (5.2.16) in elementary functions is not calculated.

The density of probability of molecules distribution in the velocities, using (5.2.15), looks like:

$$\frac{dn}{ndv} = \frac{F}{n}\frac{v^2}{1+\exp\left(\frac{mv^2/2+\Delta\varepsilon}{kT}\right)}. \tag{5.2.17}$$

The value is a change of molecules' energy at the transition of electrons from level to level depending on a kind of gas. For rarefied gas in atomic conditions, the spectrum of radiation is lined. Also, on any length of wave at a certain temperature of gas the maximal energy is radiated. For example, in a spectrum of seen light at steam of mercury (in mercury-quartz lamps) the brightest green line has a length of a wave $\lambda = 546.1\ nm$. Strictly speaking, the value is necessary to calculate under the formula where the summation is executed on all possible energies' transitions in a spectrum of gas. For further analysis we shall assume that the radiation of rarefied monochromatic gas occurs on one length of the wave corresponding to a maximum of energy of thermal equilibrium radiation of black body. The energy of the electrons' transition at which the maximum of energy of thermal radiation is observed, is calculated from a ratio (Wien's law) $\Delta\varepsilon = h\nu = 4.96kT$, [21].

Let's designate in (5.2.16) the value $\frac{mv^2}{2kT} = X$. The resulting formula (5.2.16) can be transformed:

$$n = F\frac{1}{2}\left(\frac{2kT}{m}\right)^{\frac{3}{2}}\int_0^{\infty}\frac{\sqrt{X}dX}{1+\exp(X+4.96)} = F\frac{1}{2}\left(\frac{2kT}{m}\right)^{\frac{3}{2}}6.131\cdot10^{-3}. \tag{5.2.18}$$

In the formula (5.2.18) and lower all integrals are found by the numerical method.

Hence, the value

$$F = 4\pi a\left(\frac{m}{2\pi kT}\right)^{\frac{3}{2}}n, \tag{5.2.19}$$

where constant $a = 144.5$. Size a and number π are used for comparison of distribution (5.2.17) with Maxwell's distribution [37]:

$$\frac{dn}{ndv} = 4\pi\left(\frac{m}{2\pi kT}\right)^{\frac{3}{2}} v^2 \exp\left(-\frac{mv^2/2}{kT}\right). \tag{5.2.20}$$

Substituting (5.2.19) in (5.2.17), we find:

$$\frac{dn}{ndv} = 4\pi a\left(\frac{m}{2\pi kT}\right)^{\frac{3}{2}} \frac{v^2}{1+\exp\left(\frac{mv^2/2+\Delta\varepsilon}{kT}\right)}. \tag{5.2.21}$$

Distribution (5.2.21) actually depends on three variables: v, $\Delta\varepsilon$, and T.

Let's now consider a case of non-excited molecules when the function of distribution has the form (5.2.8).

Similarly (5.2.15), we shall write down:

$$dn = F\frac{v^2 dv}{\exp\left(\frac{mv^2/2+\Delta\varepsilon}{kT}\right)-1}. \tag{5.2.22}$$

As before, we accept $\Delta\varepsilon = 4.96kT$. Constant F we shall find from the condition of normalization:

$$n = F\int_0^\infty \frac{v^2 dv}{\exp\left(\frac{mv^2/2+\Delta\varepsilon}{kT}\right)-1}. \tag{5.2.23}$$

Hence:

$$F = 4\pi b\left(\frac{m}{2\pi kT}\right)^{\frac{3}{2}} n, \tag{5.2.24}$$

where $b = 143.8$.

Thus, the distribution of non-excited molecules looks like:

$$\frac{dn}{ndv} = 4\pi b\left(\frac{m}{2\pi kT}\right)^{\frac{3}{2}} \frac{v^2}{\exp\left(\frac{mv^2/2+\Delta\varepsilon}{kT}\right)-1}. \tag{5.2.25}$$

In figure 5.2.5 the curves of functions of distribution, which are described by formulas (5.2.21), (5.2.25) - curve 1, are shown, and Maxwell's distribution (5.2.20) - curve 2. The temperature of gas (air with equivalent molecular mass 29) is accepted as equal $T = 300\,K$.

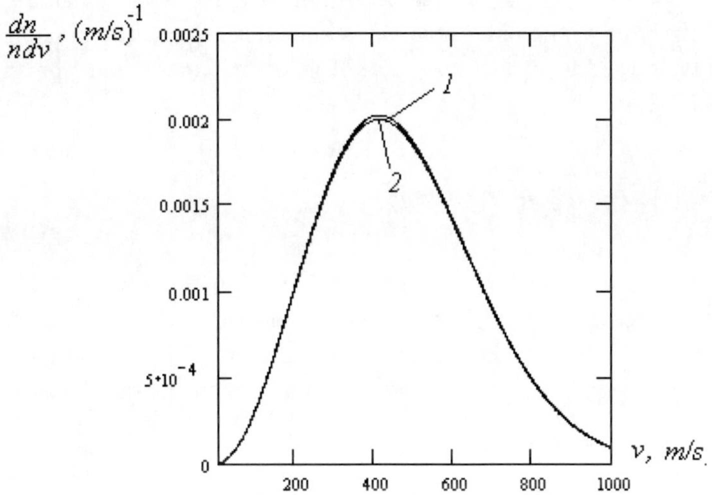

Figure 5.2.5.

From the figure it is apparent that the difference between distributions (5.2.21) and (5.2.25) is practically absent. The distinction between them and Maxwell's distribution does not exceed 1%. The curve of Maxwell's distribution passes a little below the curves plotted under formulas (5.2.21) and (5.2.25) only in the areas of the most probable velocity of molecules.

Hence, for the calculation of speeds of molecules and other statistical parameters of molecular radiating systems with the biggest accuracy, it is possible to use the Maxwell distribution.

A comparison of distributions (5.2.21) and (5.2.25) and Maxwell's distributions shows that Maxwell's distribution (5.2.20) is an average distribution between (5.2.21) and (5.2.25).

That distributions of velocities of the excited and non-excited molecules (5.2.21) and (5.2.25) are numerically practically identical is defined by the action of the law of quantity movement conservation of molecules before and after collision, which for differentials looks like:

$$\boldsymbol{dp} + \boldsymbol{dp}_1 = \boldsymbol{dp}_2 + \boldsymbol{dp}_3. \tag{5.2.26}$$

Concerning the law of conservation (5.2.26), it is necessary to make the following remark. It is applicable and appropriate for a situation when, in the condition of balance of the system, the left part reflects the quantity of movement of particles leaving some phase volume at collision, and the right part reflects the quantity of movement of the particles passing at collision in this phase volume. The same can be attributed to the law of conservation of energy (5.2.12).

From (5.2.26) we can deduce that:

$$(dp)^2 + (dp_1)^2 + 2d\boldsymbol{p}d\boldsymbol{p}_1 = (dp_2)^2 + (dp_3)^2 + 2d\boldsymbol{p}_2 d\boldsymbol{p}_3 . \tag{5.2.27}$$

At the interaction of molecules in the condition of balance, the quantity of molecules leaving and passing in the same phase volume is kept. Therefore, it is possible to write that [21]. This equality is used at deduction of the Boltzmann kinetic equation (5.2.5). Hence, from (5.2.27) we have:

$$(dv)^2 + (dv_1)^2 = (dv_2)^2 + (dv_3)^2 . \tag{5.2.28}$$

The ratio (5.2.28) shows that a change of velocities of molecules 2 and 3 can occur only owing to a change of velocities of molecules without index and 1, but not due to radiation by the excited molecules 2 and 3 of electromagnetic quantums. It also determines the similarity (figure 5.2.5) of distributions (5.2.21) and (5.2.25), despite their various forms.

5.2.4. Energy Aspects of Rarefied Gas Radiation

Let's find a share of energy which passes on radiation of gas.

The general energy of n gas molecules can be found under the formula:

$$U_{\pm} = \int_0^{\infty} W dn_{\pm} , \tag{5.2.29}$$

where the top sign is attributed to the excited molecules, and the bottom sign to the non-excited.

Using formulas (5.2.15) and (5.2.22), we shall find:

$$U_{\pm} = F_{\pm} \int_0^{\infty} \frac{\left(mv^2/2 + \Delta\varepsilon\right)v^2}{\exp\left(\dfrac{mv^2/2 + \Delta\varepsilon}{kT}\right) \pm 1} dv . \tag{5.2.30}$$

Carrying out separate integration for the excited and non-excited molecules, we have:

$$U_+ = \frac{2a}{\sqrt{\pi}} nkT \int_0^\infty \frac{(X+4.96)X^{1/2}}{\exp(X+4.96)+1} dX = 6.52nkT, \quad (5.2.31)$$

$$U_- = \frac{2b}{\sqrt{\pi}} nkT \int_0^\infty \frac{(X+4.96)X^{1/2}}{\exp(X+4.96)-1} dX = 6.517nkT. \quad (5.2.32)$$

As before, the substitution $\frac{mv^2}{2kT} = X$ is made. Besides the sizes of constants *F* from formulas (5.2.19) and (5.2.24) are used.

From (5.2.31) and (5.2.32), it follows that the energy of radiating gas far exceeds the internal energy of ideal gas $U = 1.5nkT$. It is connected by the fact that the greatest fraction of energy *Q* passing in the gas system transits on the excitation of molecules. In the condition of balance the isovolumetric gas system is examined, figure 5.2.4, on the radiation of gas is spent energy:

$$\Delta U = Q = U_+ - U_- = 3.2 \cdot 10^{-3} nkT. \quad (5.2.33)$$

Let's find, following [38], the energy of thermal radiation falling 1 quantum. Using Bose-Einstein's distribution for photon gas [37], we shall find the volumetric density of energy of radiation:

$$\varepsilon = \frac{2}{(2\pi\hbar)^3} \int_0^\infty \frac{cp \cdot 4\pi p^2 dp}{\exp(ap)-1} = \frac{8\pi c}{(2\pi\hbar)^3} \frac{\pi^4}{15a^4}, \quad (5.2.34)$$

where *p* is the impulse of quantum; also, it is designated $a = \frac{c}{kT}$.

The number of quantums in a unit of volume equally:

$$n_0 = \frac{2}{(2\pi\hbar)^3} \int_0^\infty \frac{4\pi p^2 dp}{\exp(ap)-1} = \frac{8\pi}{(2\pi\hbar)^3} \frac{2\varsigma(3)}{a^3}, \quad (5.2.35)$$

where $\zeta(3)$ is the zeta function of Riemann.

The average energy of thermal radiation falling 1 quantum equals:

$$\varepsilon_1 = \frac{\varepsilon}{n_0} = \frac{\pi^4}{30\varsigma(3)} \frac{c}{a} = \frac{\pi^4}{30\varsigma(3)} kT \approx 2.7kT. \quad (5.2.36)$$

Let's find the quantity of simultaneously radiating molecules $N = \dfrac{\Delta U}{\varepsilon_1} = 1.18 \cdot 10^{-3} n$. Hence, they simultaneously radiate only all molecules of gas.

If we are to consider rarefied gas in the given thermodynamic system as the converter of thermal energy in the energy of electromagnetic radiation on the basis of the carried out analysis, it is possible to make the following conclusion. The energy of transition from the thermal form, for example, by heating a vessel with gas, in energy of thermal radiation is a very small part of the general internal energy of gas. That gas practically submits to Maxwell's distribution. If we are to use conditional efficiency of the transformation of thermal energy into energy of radiation for the gas system it makes:

$$\eta = \frac{\Delta U}{U_+} 100\% \approx 0.05\%. \tag{5.2.37}$$

In spite of the fact that transitions of molecules at their collisions from one form of distribution into another do not practically influence the distribution of molecules' velocities, these transitions determine the distribution of quantums of thermal radiation on their energies.

5.3. Kinetic Processes in Electronic Gas of Metal

Let's consider some kinetic processes in the electronic gas of metal. An interesting distinctive feature of metal is that at low temperatures it transits in a superconducting state.

The observable experimental linear dependence of the specific resistance of metal and its absolute temperature is usually correct in rather wide interval of temperatures. But at low temperature there is a sharp nonlinear decrease in specific resistance ρ and at the achievement of some critical temperature T_c, resistance of metal practically disappears. Metal passes in a superconducting condition. The phenomenon of superconductivity was revealed by Dutch physicist H. Kamerlingh-Onnes in 1911. It was in sharp decrease practically up to a zero level of resistance of mercury at temperatures close to zero degrees of Kelvin. Now the theory of a superconducting state is developed in enough phenomenological detail (L.D. Landau, V.L. Ginzburg), and with quantum the points of view (J. Bardeen, L. Cooper, J. Schrieffer).

From the physical point of view the superconducting state arises owing to the existence of energy split in zoned structure of substance owing to the fact that the electrons moving under the action of an electric field at temperatures below critical cannot transfer the energy to a crystal lattice. But in addition to this, undoubtedly, to the true position, it is frequently supposed that at temperatures below critical, and only in this case, electrons are united in the so-called Cooper's pairs.

In the basis of the quantum theory of superconductivity is the assumption of special carriers of charges in substance – Cooper's pairs. The Cooper's pair will consist of two coupled electrons with opposite directions of spins. As total spin it equals zero from the point of view of quantum statistics; the Cooper's pair is boson. As opposed to electrons, which are fermions, Cooper's pairs do not submit to the Pauli principle of veto and can be in the

identical energy state. If the temperature of a conductor is close to zero all Cooper's pairs occupy the identical lowest energy state.

In case Cooper's pairs cannot transfer energy of the directed movement to ions of crystal lattice, resistance of the conductor becomes equal to zero. The impossibility of transferring energy from Cooper's pairs to the crystal lattice specifies, from the point of view of the zoned energy theory, the existence of forbidden zone Δ in a quantum energy spectrum of crystal lattice which exceeds the energy of Cooper's pairs at the lowest energy level.

The assumption that Cooper's pairs arise only at temperatures below critical is not obvious. At temperatures below the critical point, undoubtedly, all electrons are united in Cooper's pairs. But there is no veto on the time existence of Cooper's pairs and at temperatures higher than critical.

Let's consider in more detail the question of the potential for the existence of Cooper's pairs in metal at temperatures higher temperatures of transition in a superconducting state. For example, in [40] this question was thoroughly analyzed in detail; therefore, there is no necessity to repeat it completely.

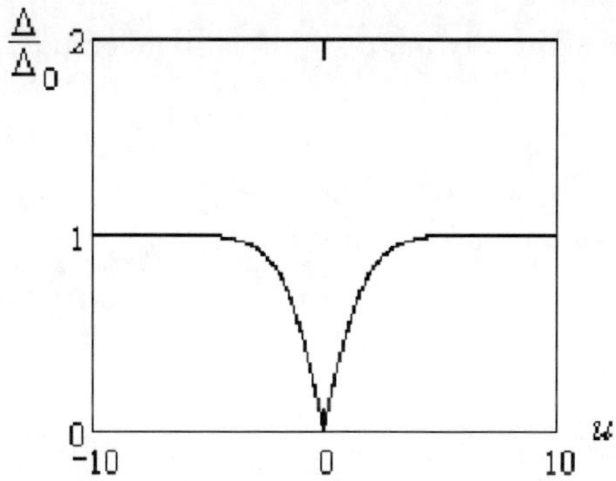

Figure 5.3.1.

A lack of the given analysis is the approximated calculation of integral $I(u)=\int_0^\infty \frac{dx}{\sqrt{x^2+u^2}\exp\left(\sqrt{x^2+u^2}+1\right)}$, where $u=\frac{\Delta}{kT}$, value Δ is half of the energy of a couple of Cooper's pairs, k – Boltzman constant. Designations [10] are retained. Furthermore, there are two members of expansion into the series of subintegral expression and the conclusion is that $\Delta \sim \sqrt{1-\frac{T}{T_c}}$, hence, the existence of Cooper's pairs at temperatures greater than critical is impossible.

The ratio between value Δ and the examined integral can be written down as [40] $\frac{\Delta}{\Delta_0}=\exp(-2I)$, where is the size of value Δ at absolute temperature $T=0$. The numerical calculation under the given formula is shown on figure 5.3.1. Negative values u, apparently, have no physical sense.

The calculating curve at is completely smooth. There are no indications of the existence of critical effects at any certain temperatures.

The only limitation which follows from the given calculation is $\frac{\Delta}{\Delta_0} \leq 1$. Close this limitation is more strict. The received result is natural enough: the energy of a couple of Cooper's pairs at any temperature is less than at $T = 0$. Any basic restrictions on the temperature at which Cooper's pairs are formed does not exist.

The formula is connected to the way of finding the integral by expansion into the series of subintegral expression and restriction by two members, and then even the first order on [40]. Therefore, at the approach of metal temperature $T \to T_c$, this formula becomes inapplicable.

For example, in the first order probably expansion into series $\sqrt{e^{-x}} \approx \sqrt{1-x}$. In the left part of the equality function has no restrictions at *x*. In the right part there is a restriction $x \leq 1$. If the physical law submits the function at expansion into the series in the first order there is false restriction. Approximately there is critical temperature, for Cooper's pairs have appeared.

We can notice that similar incidents sometimes arise in physics. For example, in many courses of general physics Thomson's formula for the pressure of saturated steam at height *h* above the curvilinear surface of a liquid [37] results: $P_h = P_0 - \frac{2\sigma}{R}\frac{\rho_S}{(\rho_L - \rho_S)}$. In the formula и is density steam and liquids, σ - factor of surface tension, *R* - radius of meniscus. At some parameters the pressure of saturated steam becomes negative; that is impossible. Actually, such restriction is incorrect, since in the right part there is an expansion into the series of exponent [21]. Actually, the dependence on the factor of surface tension σ is not linear, and exponential. Frequently, this question is not discussed in textbooks.

The purpose of the given chapter is the analysis of electrons' translation in metal under the action of electric field in the absence of temperature restrictions on the occurrence of Cooper's pairs.

5.3.1. Two Kinds of Charge Carriers in Metal Superconductivity

In the existing theory of superconductivity the general density of the current on a superconducting conductor is accepted as the normal sum connected with Joulian heat liberation, and superconducting currents [40]:

$$j = j_n + j_s . \qquad (5.3.1)$$

Let's carry out the analysis of the Ohm law in metal on the assumption that there is the probability of the existence of Cooper's pairs at higher temperatures of transition in a superconducting state T_c.

Let the conductor be a homogeneous metal film on a surface of dielectric on which the direct current flows under the action of the electric field, figure 5.3.2.

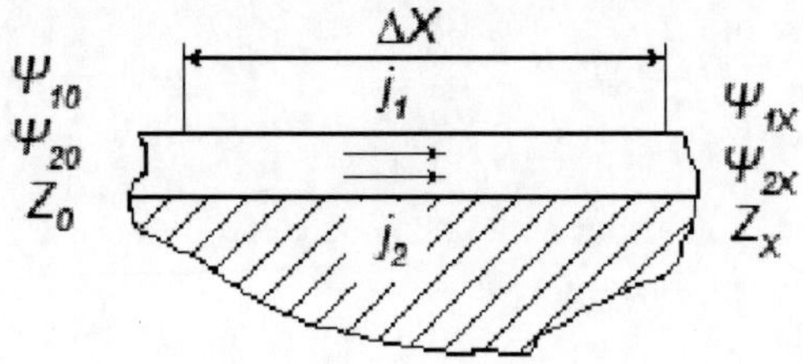

Figure 5.3.2.

The thickness of the conducting film is small, so the temperature across a conductor is identical. We shall allocate the length of the conductor ΔX .

Let's assume that the temperature of the conductor is not ultralow, but also that in it Cooper's pairs are formed. Concentration of these pairs is defined by the dynamic balance of their continuous disintegration and formation at the given temperature. Cooper's pairs transfer current alongside with separate free electrons. Thus, the current in the conductor both fermions and bosons are transferred. Therefore, for our purposes it is more convenient to present the general density of a current in the conductor as:

$$j = j_1 + j_2, \tag{5.3.2}$$

where index 1 is attributed to the charge carrier - electron, and index 2 to boson, figure 5.3.2.

The Ohm law for such a conductor can be written down as:

$$j = -\frac{1}{\rho_1}\frac{\partial \varphi_1}{\partial X} - \frac{1}{\rho_2}\frac{\partial \varphi_2}{\partial X}. \tag{5.3.3}$$

In the formula (5.3.3) and are the specific resistance caused, accordingly, in electrons and Cooper's pairs, X - coordinate along the conductor.

The equation (5.3.3) should be added with the balance of potential increments:

$$\Delta\varphi = \Delta\varphi_1 + \Delta\varphi_2\,, \tag{5.3.4}$$

where is a potential increment on conductor length ΔX , - a part of the potential increment, causing electronic current, - a part of the potential increment, causing Cooper's current. We shall note that the potential increment has a sign opposite to the potential difference.

Let's transform the equation (5.3.3) to a kind:

$$j\rho_2 + \psi\frac{\partial \varphi_1}{\partial X} = -\frac{\partial \varphi_2}{\partial X}, \tag{5.3.5}$$

where is the ratio of Cooper's and electronic specific resistances.

Let's designate:

$$Z = j\rho_2 + \psi \frac{\partial \varphi_1}{\partial X} = j\rho_2 + \psi \frac{\Delta \varphi_1}{\Delta X}, \qquad (5.3.6)$$

where in the formula (5.3.6) linear dependence is accepted.

Hence, the equation (5.3.5) becomes

$$Z = -\frac{\partial \varphi_2}{\partial X}. \qquad (5.3.7)$$

Solving (5.3.7) by a method of separation of variables, we find:

$$\int_0^{\Delta X} dX = -\int_{\varphi_{20}}^{\varphi_{2X}} \frac{d\varphi_2}{Z}, \qquad (5.3.8)$$

where is Cooper's potential on the left end of the conductor section, - Cooper's potential on the right end of conductor section in length ΔX, figure 5.3.2.

Let's transit in the equation (5.3.8) to variable Z. For this purpose, using (5.3.6) and taking into account $j\rho_2 \neq f(X)$, we shall find the differential:

$$dZ = d\left(\psi \frac{\partial \varphi_1}{\partial X}\right) = \frac{\psi}{\Delta X} d(\Delta \varphi_1). \qquad (5.3.9)$$

At the finding (5.3.9) we have assumed, that the temperature of conductor T on length does not change. The value is the function of the conductor temperature. Really, the specific resistance of the conductor as a first approximation is in inverse proportion to the numerical concentration of the charge carriers [27], i.e. quantity of charge carriers in unit of volume of the conductor. Hence $\rho_1 \sim \frac{1}{N_1}$, and $\psi = \frac{\rho_2}{\rho_1} = \frac{N_1}{N_2}$, where и is the numerical concentration of electrons and Cooper's pairs. At the temperature decrease of the conductor the numerical concentration of the free electrons carrying the charge in the conductor is reduced due to the increase in concentration of Cooper's pairs. The probability of destruction of Cooper's pair is proportional; $\exp\left(-\frac{W_c}{kT}\right)$, where is energy coupling of electrons in Cooper's pair [41]. Hence, the probability of pair occurrence is:

$$\beta(T)=\frac{N_2}{N_1+N_2}=1-\exp\left(-\frac{W_c}{kT}\right) \text{ and}$$

$$\psi(T)=\frac{N_1}{N_2}=\left(\exp\left(\frac{W_c}{kT}\right)-1\right)^{-1}. \tag{5.3.10}$$

It is obviously $\psi(T)=\frac{1}{\beta(T)}-1$. At the probability of occurrence of Cooper's pairs $\beta\to 1$, the parameter $\psi(T)\to 0$.

Let's assume that at temperature all free electrons are united in Cooper's pairs, i.e. and $N_2\to N_{2\max}$. Therefore, it is possible to accept $\lim \psi\to 0$ at $T\to 0$.

Using (5.3.9), (5.3.4) and $\Delta\varphi = const$, we shall find:

$$dZ=-\frac{\psi}{\Delta X}d(\Delta\varphi_2)=-\frac{\psi}{\Delta X}d(\varphi_2-\varphi_{20})=-\frac{\psi}{\Delta X}d\varphi_2. \tag{5.3.11}$$

Finding from (5.3.11) and substituting this differential in the equation (5.3.8), we find:

$$\Delta X=\frac{\Delta X}{\psi}\int_{Z_0}^{Z_X}\frac{dZ}{Z} \text{ or } 1=\frac{1}{\psi}\int_{Z_0}^{Z_X}\frac{dZ}{Z}, \tag{5.3.12}$$

where and are values of function Z on borders of the conductor section in length ΔX, fig. 5.3.2. The received ratio (5.3.12) is equivalent to Cooper's condition for the existence of the coupling state of electrons [27]. Note that for the deduction of Cooper's condition, we actually did not use quantum representations.

Integrating (5.3.12), we find:

$$Z_X=Z_0\exp(\psi). \tag{5.3.13}$$

Let's transit in (5.3.13) to potentials, using (5.3.7):

$$j\rho_2+\psi\frac{\Delta\varphi_{1X}}{\Delta X}=\left(j\rho_2+\psi\frac{\Delta\varphi_{10}}{\Delta X}\right)\exp(\psi), \tag{5.3.14}$$

where and are increments of electronic potential on the conductor section in length ΔX, figure 5.3.2.

With the account (5.3.4) we have:

$$\Delta\varphi_{10} = \Delta\varphi - \Delta\varphi_{20},\ \Delta\varphi_{1X} = \Delta\varphi - \Delta\varphi_{2X}, \tag{5.3.15}$$

where and are increments of boson potential on the conductor section in length ΔX.

Substituting (5.3.15) in (5.3.14), we shall find from the equation of (5.3.14) the general current density in the conductor:

$$j = -\frac{\psi}{\rho_2}\frac{\Delta\varphi}{\Delta X} + \frac{\psi}{\rho_2 \Delta X}\left[\Delta\varphi_{20} - \frac{\Delta\varphi_{2X} - \Delta\varphi_{20}}{\exp(\psi) - 1}\right]. \tag{5.3.16}$$

The formula (5.3.16) shows that at $\psi \to 0$, i.e. at $T \to 0$, with the account $\lim \frac{\psi}{\exp(\psi) - 1} = 1$, some current density is kept:

$$j_{s0} = -\frac{\Delta\varphi_{2X} - \Delta\varphi_{20}}{\rho_2 \Delta X}, \tag{5.3.17}$$

where the index 0 in this case means that it is the current of the Cooper's pairs at temperature $T = 0$. The found density of the current is provided with only Cooper's pairs.

Let's designate addends in the equation (5.3.16):

$$j_3 = -\frac{\psi}{\rho_2}\frac{\Delta\varphi}{\Delta X} = -\frac{1}{\rho_1}\frac{\Delta\varphi}{\Delta X} = -\frac{1}{\rho_1}\frac{\Delta\varphi_1}{\Delta X} - \frac{1}{\rho_1}\frac{\Delta\varphi_2}{\Delta X}, \tag{5.3.18}$$

$$j_s = \frac{\psi}{\rho_2 \Delta X}\left[\Delta\varphi_{20} - \frac{\Delta\varphi_{2X} - \Delta\varphi_{20}}{\exp(\psi) - 1}\right]. \tag{5.3.19}$$

The current density is not connected to the phenomenon of superconductivity. It is defined by the carriage of fermions and bosons at high enough temperatures. Thus, resistance to carrying Cooper's pairs in substance depends only on resistance to the carriage of separate electrons making a pair.

More interesting is addend j_s. Taking into account (5.3.17), it is possible to identify the current with a current of superconductivity which is carried out only for the account of the Cooper's pairs. Occurrence of the potential difference, providing movement of Cooper's pairs, is connected with the association of electrons in the pair; therefore, it is possible to write down:

$$\Delta\varphi_{20} = \frac{W_{cв0}}{2e}, \qquad \Delta\varphi_{2X} = \frac{W_{cвX}}{2e}, \tag{5.3.20}$$

where and there is energy of coupling of electrons in Cooper's pairs in the beginning and the end of the conductor section in length ΔX , - a charge of the Cooper's pairs. In spite of the fact that at the left in (5.3.20) increments of potentials, instead of their differences, the sign before energy of coupling there is positive, since in the denominator the charge of pair is negative.

Substituting (5.3.20) in (5.3.19) and in (5.3.17), we have:

$$j_s = \frac{\psi}{2e\rho_2 \Delta X}\left[W_{c\text{в}0} - \frac{W_{cX} - W_{c0}}{\exp(\psi) - 1}\right] \text{ and}$$

$$j_{s0} = -\frac{W_{cX} - W_{c0}}{2e\rho_2 \Delta X}. \qquad (5.3.21)$$

At the deduction (5.3.16) - (5.3.21), despite the application of quantum terminology (fermions, bosons, etc.), quantum representations about the charge carriers in the conductor have practically not been used. Therefore, with the help of the analysis it is impossible to find the size of the critical temperature of the conductor transition in superconducting state. However, there is the indication that such temperature exists. Really, at a decrease in temperature from critical size up to zero and, accordingly $\psi(T_c) \to \psi(T) \to 0$, there is an increase of a current from zero up to $j_s = j_{s0}$, i.e. $0 \to j_s \to j_{s0}$.

The condition corresponds to the critical size of temperature (it will be marked in paragraph 5.3.2), i.e. according to (5.3.21) or $\psi(T_c) = \ln\frac{W_{cX}}{W_{c0}} = \ln\eta$, where $\eta = \frac{W_{cX}}{W_{c0}}$. Unfortunately, the received condition does not allow us to calculate the numerical size of the critical temperature of substance T_c

In figure 5.3.3 the dependence of the relative superconducting current $\frac{j_s}{\langle j\rangle} = \psi\left[1 - \frac{\eta - 1}{\exp(\psi) - 1}\right]$ from ψ for size $\eta = \exp[\psi(T_c)] = 1.14$, where $\langle j\rangle = \frac{W_{c0}}{2e\rho_2 \Delta X}$ is shown. Thus, according to (5.3.10), the size is $\psi(T_c) = \frac{1}{\beta(T_c)} - 1 = 0.132$ where $\beta(T_c) = 0.883$ for lead it will be found below in paragraph 5.3.2, figure 5.3.6.

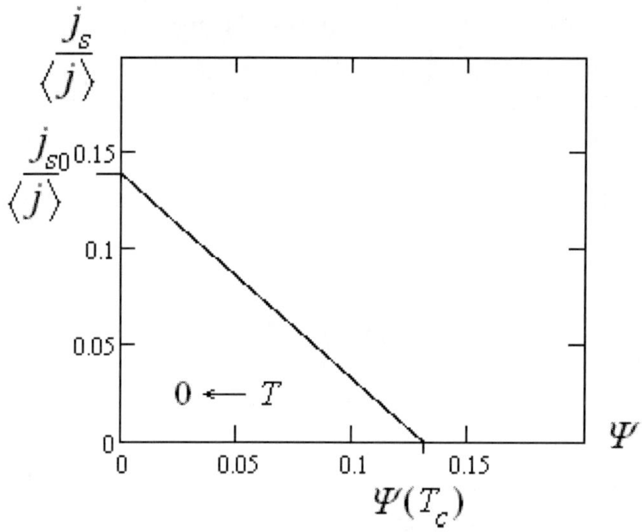

Figure 5.3.3.

In the figure 5.3.3 it is taken into account due to the negative charge of the electron in the denominator.

5.3.2. Quantum Features of Charge Carry in Conductor

To show that the received density of the current (5.3.17) is the current of superconductivity, it is necessary to address quantum representations.

Let's consider the zoned energy structure of substance of the conductor at the lowest energy levels; i.e. at ultralow temperatures, figure 5.3.4.

As already specified earlier, the existence of energy split Δ is typical of substance in this area. Energy of coupling separates energy of disintegration of Cooper's pair from energy of coupled state *E*, so on initial of the conductor section [4]:

$$\varepsilon_K + \varepsilon_0 - E = W_{c0}, \tag{5.3.22}$$

where there is energy of the Fermi level, - energy of system of two non-coupling electrons above Fermi's level.

After passage of distance to the conductor, energy of coupling of Cooper's pair has changed, figure 5.3.4:

$$\varepsilon_K + \varepsilon_X - E = W_{cX}, \tag{5.3.23}$$

where there is energy of two non-coupled electrons before the creation of Cooper's pair on the end of the conductor section.

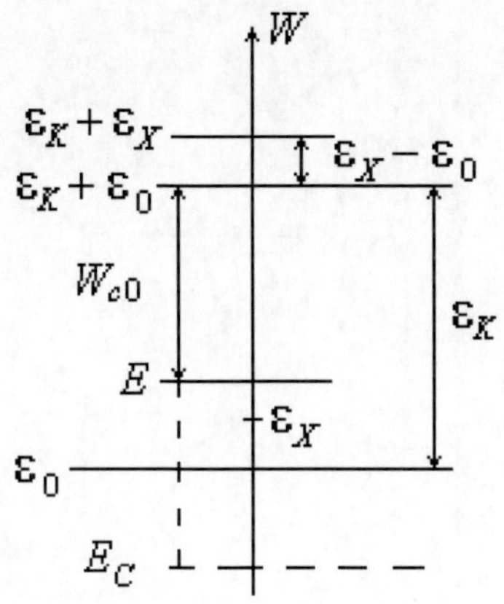

Figure 5.3.4.

Substituting (5.3.22) and (5.3.23) in (5.3.21), we find a superconducting current of Cooper's pairs as:

$$j_s = \frac{\psi}{2e\rho_2 \Delta X}\left[\varepsilon_K + \varepsilon_0 - E - \frac{\varepsilon_X - \varepsilon_0}{\exp(\psi) - 1}\right]. \tag{5.3.24}$$

If and $W_{cX} > \Delta$, i.e. on all the conductor section the energy of coupling exceeds the size of energy split, the superconductivity does not come since the energy of coupling can be transferred to crystal lattice. If and $W_{cX} \le \Delta$, on the conductor section there is a superconducting state since the energy of coupling cannot be transferred to crystal lattice. The Cooper's pairs are kept on the conductor section ΔX. This phenomenon is observed at a decrease in temperature of the conductor to critical size $T \to T_c$. At critical temperature a ratio between energy of coupling and width of energy split and $W_{cX} = \Delta$. Thus a current $j_s = 0$. The further decrease in the temperature of substance below critical results in an increase in the superconducting current j_s. Thus, at (and $T = T_c$) it is possible, proceeding from (5.3.24), to write down:

$$W_{c0} = \varepsilon_K + \varepsilon_0 - E = \frac{\varepsilon_X - \varepsilon_0}{\exp(\psi) - 1} = \Delta. \tag{5.3.25}$$

A similar result is found in [4] on the basis of the solution of Schrodinger's equation for Cooper's pairs in impulse representation.

During the increase in density in Cooper's pairs at a decrease of temperature $T \to T_c$, size $\varepsilon_K \to 0$. The successful graphic illustration of the given process, is present, for example, in [4], figure 5.3.5.

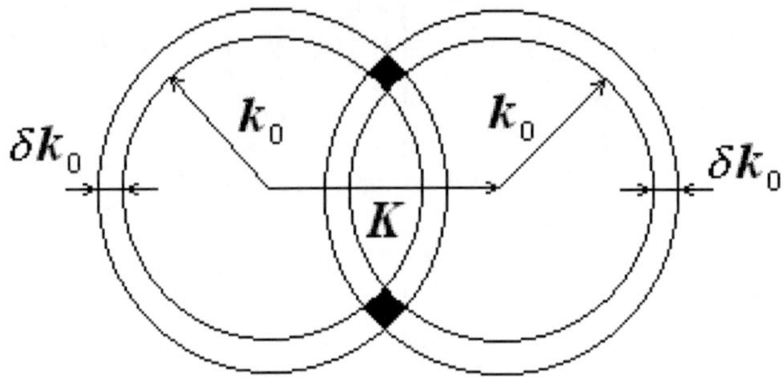

Figure 5.3.5.

In figure 5.3.5 the process of electrons associated in Cooper's pair in area of wave numbers is shown. The total impulse of Cooper's pair is proportional to the vector of the wave number connecting the centers of spheres. The impulse of the electron is connected to wave number by ratio $\mathbf{p} = \hbar \mathbf{k}$, where is Planck's reduced constant. Vectors characterize Fermi's spheres of two electrons. Impulses of separate electrons lay in areas above Fermi's level. Black areas of spheres interacting there are proportional to the density of Cooper's pairs in the conductor. From fig. 5.3.5 it is visible that the Cooper's pairs' density achieves maximum at coincidence spheres, i.e. as was noted earlier, at temperature $T \to T_c$. It results in the size of a total impulse that is equivalent to case $\varepsilon_K = 0$.

In figure 5.3.4 the dotted line shows case when during the association of the increasing electrons' number in Cooper's pairs, i.e. at $\varepsilon_K \to 0$, energy of the coupled state falls below Fermi's level due to big energy split Δ:

$$W_{c0} = \varepsilon_K + \varepsilon_0 - E = \varepsilon_0 - E_C \leq \Delta. \tag{5.3.26}$$

In this case the Cooper's pair at the movement cannot transfer energy to ions of crystal lattice and there is the superconducting state.

Let's plot the graph of the fraction of Cooper's pairs in the general number of free charge carriers in metal by the example of lead *Pb* from its temperature under the formula (5.3.10):

$$\beta = \frac{N_2}{N_1 + N_2} = 1 - \exp\left(-\frac{W_c}{kT}\right). \tag{5.3.27}$$

The width of forbidden zone Δ at a temperature close to absolute zero can be found from the ratio:

$$\frac{\Delta}{kT_c} = \xi, \tag{5.3.28}$$

where for lead size [27].

For critical temperature of transition of lead in the superconducting state we shall find the size of the energy split $\Delta = 21.3 \cdot 10^{-23}\ J$.

Superconductivity, as was already specified, will arise in the case of the energy coupling of Cooper's pairs being equal to the width of the forbidden zone. In this case, see (5.3.25), correct energy ratio:

$$W_c = \varepsilon_K + \varepsilon_0 - E = \Delta . \quad (5.3.29)$$

Assuming that the width of the forbidden zone of metal crystal lattice does not strongly depend on temperature, according to (5.3.27) it is possible to write down:

$$\beta = 1 - \exp\left(-\frac{\Delta}{kT}\right). \quad (5.3.30)$$

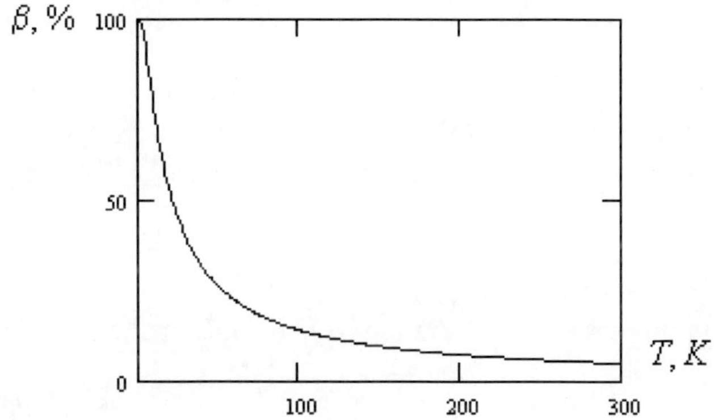

Figure 5.3.6.

In figure 5.3.6 the dependence of value β in percentage of temperature for lead is shown. As follows from the curve, at room temperature ($T = 300\ K$) the Cooper's pairs make about 5 % of all charge carriers. At the temperature of transition of lead in superconducting state their calculating share grows up to ($\beta(T_c) = 0.883$). As was already specified, the ratio (5.3.10), on the basis of which the formula (5.3.30) is found, is received in the absence of the quantum approach. Therefore, in the formula (5.3.30) there is no parameter for critical temperature T_c. Really, after the achievement of critical temperature the share of Cooper's pairs is equal to 100 %.

Let's find the interrelation of parameters of the examined model and parameters of the theory of Ginzburg - Landau [40]. It is natural to write that the current of Cooper's pairs there is:

$$j_2 = j_s = \frac{1}{\rho_2}\frac{\partial \varphi_2}{\partial X} = \frac{eh}{m}|\psi|^2\frac{\partial \Phi}{\partial X}, \tag{5.3.31}$$

where e is the electron charge, m - its mass, - square of the module of parameter of the order, Φ - its phase, so $\psi = |\psi|\exp(i\Phi)$, and $|\psi|^2 = N_2$. It is taken into account that in a Cooper's pair there are two electrons.

Using (5.3.31) it is possible to write down:

$$\Delta\varphi_2 = \frac{eh\rho_2}{m}|\psi|^2\Delta\Phi = \frac{eh\rho_2}{m}N_2\Delta\Phi = \theta h\Delta\Phi, \tag{5.3.32}$$

where it is designated $\theta = \frac{e\rho_2}{m}N_2$.

Taking into account the phase change equally where n is integer [27, 40], we shall obtain:

$$\Delta\varphi_2 = 2\pi h\theta n. \tag{5.3.33}$$

Thus, a part of the potential increment, causing Cooper's current, is quantized. The minimal size of a potential increment there is $n = 1$, i.e. $\Delta\varphi_{2\min} = 2\pi h\theta$.

Let's estimate the minimal size $\Delta\varphi_{2\min}$. According to Drude's theory [40] complex $\rho_2 N_2 = \frac{m}{2e^2\tau}$. In this case, the value τ can be treated as the time of Cooper's pair life at the given temperature. According to the Frenkel kinetic theory of liquid [42], for Bose-condensate of Cooper's pairs it is possible to write down:

$$\tau = \tau_0 \exp\left(\frac{W_c}{kT}\right), \tag{5.3.34}$$

where is the period of electron oscillation in the pair, - frequency of its oscillations.

Application of the formula (5.3.34) is allowable, since the real system of electrons in metal on the properties is more similar to electronic liquid than to gas [4].

Hence:

$$\rho_2 N_2 = \frac{m}{2e^2\tau} = \frac{m\nu_0}{2e^2}\exp\left(-\frac{W_c}{kT}\right). \tag{5.3.35}$$

Substituting the given complex in the formula for $\theta = \frac{e\rho_2}{m} N_2$, we obtain:

$$\theta = \frac{\nu_0}{2e} \exp\left(-\frac{W_c}{kT}\right). \tag{5.3.36}$$

Thus, at from (5.3.33) we have:

$$\Delta\varphi_{2\min} = \pi \frac{h\nu_0}{e} \exp\left(-\frac{W_c}{kT}\right). \tag{5.3.37}$$

Accepting $\Delta = W_c$, the value of the energy quantum causing Cooper's current can be calculated under the formula:

$$\Delta E_K = 2\pi h\nu_0 \exp\left(-\frac{\Delta}{kT}\right), \tag{5.3.38}$$

where the factor 2 means two electrons in Cooper's pair.

The value of the energy quantum causing Cooper's current in lead at critical temperature $T_c = 7.19\ K$, taking into account (5.3.28) and $\xi = 2.15$, equals:

$$\Delta E_K = 2\pi h\nu_0 \exp(-2.15) = 0.232\,\pi h\nu_0. \tag{5.3.39}$$

The carried out kinetic analysis of charges carried in the conductor which can transit in the superconducting condition shows that the basic laws of carriage follow from classical representations in the assumption that electrons can be united in Cooper's pairs which concentration is defined by the dynamic balance of their occurrence and disintegration. The substance of the conductor at the usual temperature shows resistance to the current of Cooper's pairs.

The quantum analysis of the process of carriage allows the identity currents (5.3.21) with superconducting currents.

The energy causing Cooper's current there is quantized. Comparison of the analysis with the Ginzburg - Landau theory allows there is an estimate of the minimal size of quantum of this energy.

Chapter 6

Quantum Electrodynamics of the Interaction of an Electromagnetic Field with a Substance

In quantum electrodynamics the analysis of interaction of electromagnetic field with substance is widely used on the basis of research of effective section of interaction. Interaction of polarized light with substance is one of the most important and widespread phenomena in optics on which we shall reflect in more detail.

Full consideration of the question should be based on quantum electrodynamics. Quantum electrodynamics is now basically based on two undoubtedly correct achievements: Dirac's equation for particles and a calculative method of Feynman's diagrams. But to quantum electrodynamics (in its modern form) there are also widely known exceptions.

First, energy of quantum oscillator does not disappear at the lowest energy level. The identification of photon with quantum oscillator results in the necessity for the introduction of vacuum energy dependent on photon frequency. This dependence remains and in the absence of photon, that is completely absurd.

Secondly, if we were to summarize the vacuum energy on the infinite number of all possible harmonics with growing frequency, this sum will be divergent. The same infinity arises at the summation of electromagnetic radiation modes.

Thirdly, Maxwell's theory, together with the relativistic electrons theory, results in the equations of their movement which can be solved by method of the theory of indignations. Solutions are expressed as power series on fine-structure constant which is small parameter. However, the received series are divergent, and the maximum addends infinite.

Except for these known exceptions to quantum electrodynamics there is an obvious flaw: there is no equation for photon similar to Dirac's equation for the particle. The remark [18] that the role of the Schrodinger equation for a photon is played the Maxwell equations, it is hardly possible to assume the serious precondition of the theory of quantum. This opinion breaks physical logic. Really, the Schrodinger equation, having the wave nature, describes wave properties of electron which is basically a particle of substance. Corpuscular properties of photon, which is basically an element of electromagnetic field, cannot be described by wave of Maxwell equations. Following a conclusion from paragraph 2.1, that physical sense of the Maxwell equations is the reflection of a continuity of electromagnetic field, it completely contradicts the basic paradigm to quantum electrodynamics – the corpuscular nature of electromagnetic radiation. The equations describing corpuscular properties of photon, apparently, should follow the second law of Newton. Moreover, by consideration of the wave properties of the electron the coordinate of its position is lost while consideration of the corpuscular properties of a photon on the contrary should result in the occurrence of its space coordinate.

Therefore, despite a line of successes of quantum electrodynamics, first of all in an explanation of Lamb's shift and the abnormal electron magnetic moment, Dirac, one of the founders of quantum electrodynamics, believed that a fundamental change of the theory was necessary.

Despite all of the aforementioned remarks, the conclusions received with the help of quantum electrodynamics substantially reflect real natural phenomena.

6.1. Polarization of Light

The concept of polarized light was entered into the science of optics for the first time by Newton in 1704.

Let's consider the basic physical phenomena connected to the polarization of light.

According to the theory of Bohr, at the transition from a higher energy level to lower, the electron of atom radiates electromagnetic quantum with the certain plane of light vector oscillations.

The vector of intensity of electric field in a light electromagnetic wave is referred to as a light vector.

The following radiated quantum already has another plane of oscillations of light vector, figure 6.1.1.

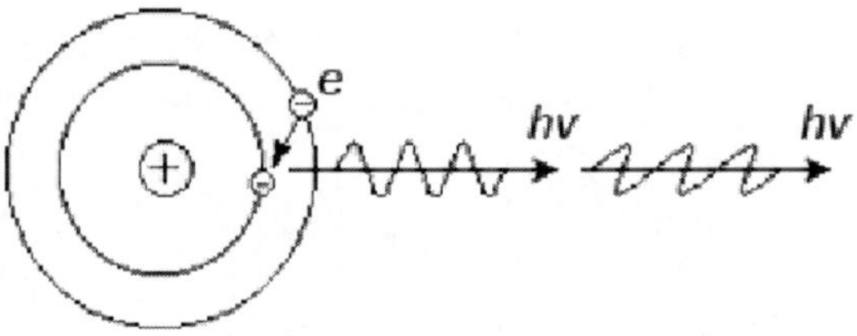

Figure 6.1.1.

The light ray is formed as a result of the addition of waves, radiated by many atoms; therefore, planes of oscillations of the light vector in light ray continuously and randomly vary.

If the plane of oscillations of the light vector randomly varies, and amplitudes of light vector in all directions are identical, such light is referred to as natural or non-polarized.

In figure 6.1.2a the section of light ray with quantums of light is conditionally shown, oscillations of which light vectors occur in various planes. There are translations of light vectors of quantums to the uniform center which are also shown.

If fluctuations of light vector occur only in one plane, such light is referred to as completely linearly polarized (or plainly polarized); figure 6.1.2b.

If, in some directions of amplitude of oscillations of light vector, it is more than in others such light is referred to as partly polarized; figure 6.1.2c.

There is a kind of polarization of light at which the light vector rotates around a direction of light spreading *Z*. The end of a vector can rotate a circle or ellipse. In this case, 'light' refers to polarized light on a circle or light with elliptic polarization; see paragraph 2.4.1.

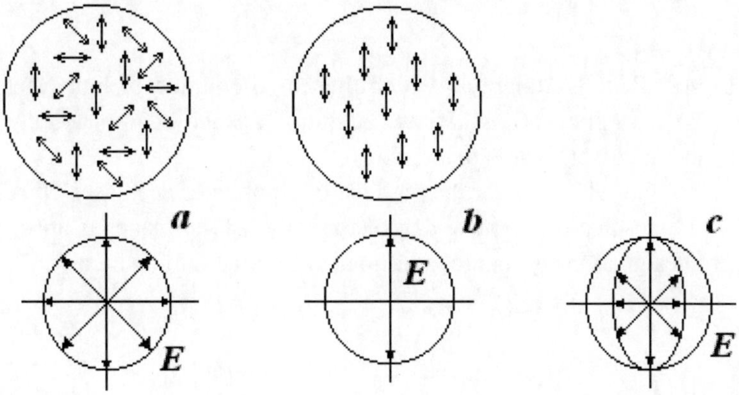

Figure 6.1.2.

In spite of the fact that in consideration of the polarization phenomenon we used the concept of light quantum, the determination of linearly polarized light is made from positions of the wave light theory. In quantum optics the polarization of light connects with the identical direction of photons spinning in the light ray.

In figure 6.1.3 the linearly polarized monochromatic wave is shown.

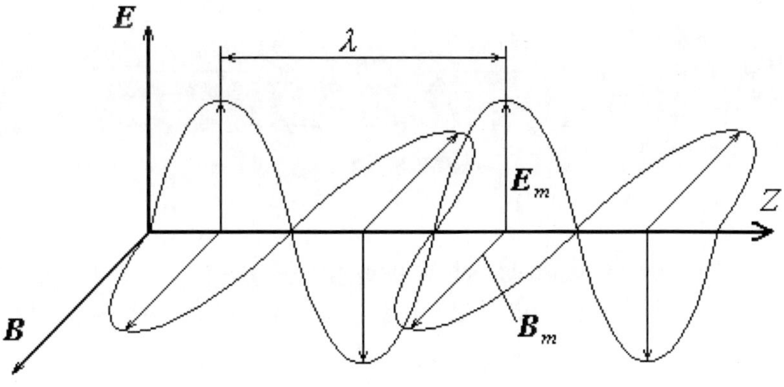

Figure 6.1.3.

On graph there is the length of a wave, vectors of electric strength and magnetic induction fields in a wave.

The device for the polarization of light is referred to as a polarizer.

Let's designate a polarizer, as shown in figure 6.1.4. The polarizer used for finding the light polarization plane is referred to as the analyzer.

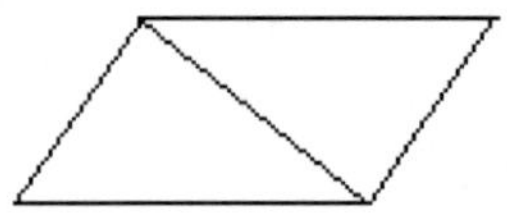

Figure 6.1.4.

The polarization plane is a plane in which there is oscillation of light vector of linearly polarized light. The main plane of a polarizer (or simply a plane of a polarizer) we shall name a plane in which the polarizer light is polarized.

The intensity of natural light after the passage of a polarizer decreases twice, fig. 6.1.5. It is connected by the fact that the light of one of the polarization planes is absorbed. In figure 6.1.5 the standard designations of natural and linearly polarized light are also shown.

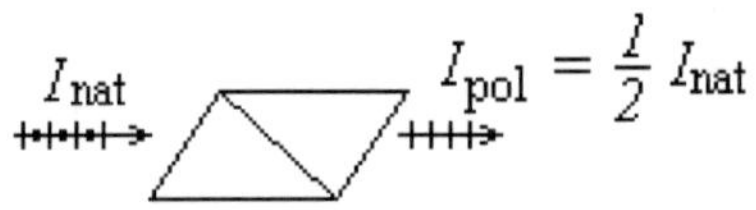

Figure 6.1.5.

6.2. MALUS' LAW. THE CLASSICAL APPROACH

Let's consider the passage of light through a polarizer system - analyzer, figure 6.2.1.

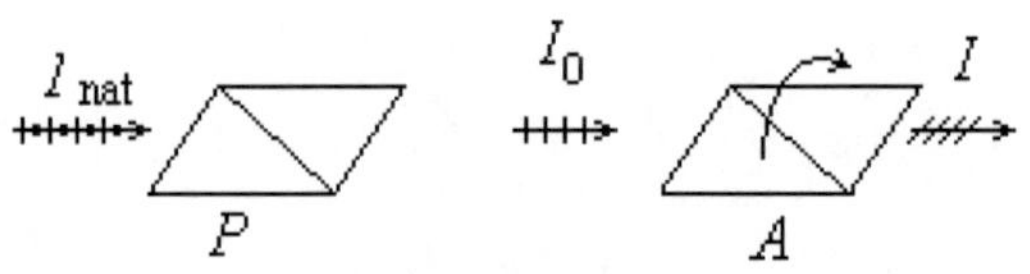

Figure 6.2.1.

The intensity of linearly polarized light leaving the polarizer *P* and falling on the analyzer *A* equals I_0. If we were to rotate the analyzer *A* about the axis the intensity of light *I* leaving the analyzer would vary. The law of light intensity *I* leaving the analyzer, depending on the angle of rotation of the analyzer plane concerning the polarizer plane, as the French scientist E.L. Malus found in 1810.

Let's say that between the polarizer and analyzer planes there is an angle α. We shall decompose a light vector (it is designated on peak sizes of vectors) of the electromagnetic wave falling on the analyzer on two components, as shown in figure 6.2.2:

$$\boldsymbol{E}_0 = \boldsymbol{E}_X + \boldsymbol{E}_Y . \qquad (6.2.1)$$

One component of light vector oscillates in the analyzer plane; the second in the perpendicular analyzer plane.

The second component will not pass through the analyzer, and the intensity of light which has left the analyzer will be defined only by the component of a light vector varying in the analyzer plane.

From figure 6.2.2 it is visible that:

$$E_X = E_0 \cos\alpha . \tag{6.2.2}$$

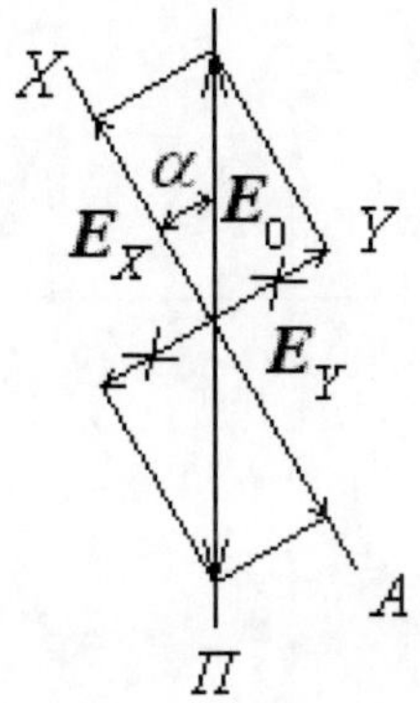

Figure 6.2.2.

The intensity of the light wave is proportional to a square of electric field strength $I \sim E_X^2$; hence:

$$I = I_0 \cos^2\alpha. \tag{6.2.3}$$

The last formula represents the contents of Malus' law.

The intensity of linearly polarized light I which has left the analyzer is proportional to the intensity of linearly polarized light falling on the analyzer and a square of cosine of the angle between planes of the analyzer and polarizer. At or intensity $I = I_0$, i.e. linearly polarized light completely passes through the analyzer. At intensity $I = 0$, i.e. light does not pass through the analyzer.

6.2.1. Polarizing Tensor. Degree of Polarization. Stokes Parameters

Considering the classical approach to the deduction of Malus' law and the problems of optical activity (see paragraph 2.4.1), it is interesting to note a more detailed description of light polarization as electromagnetic wave. In the research into optical activity, according to Fresnel's phenomenological theory (see paragraph 2.4.1), light with linear polarization was represented as the sum of two electromagnetic waves polarized on a circle (figure 2.4.2), around which light vectors rotate in different directions.

Circular polarization (figure 6.2.3) arises if, for example, the Y-component of the light vector is delayed relative to the X-components. Oscillations of a light vector for light with circular polarization can be written down as $\boldsymbol{E}(t)=\boldsymbol{E}_X e^{-i\omega t}+\boldsymbol{E}_Y e^{-i(\omega t-\varepsilon)}$, where ω there is frequency of oscillations (rotation), ε - phase of delay of oscillations *Y*-components relative *X*-components, and - as well as in the formula (6.2.1), projection of light vector on coordinates' axes, figure 6.2.3.

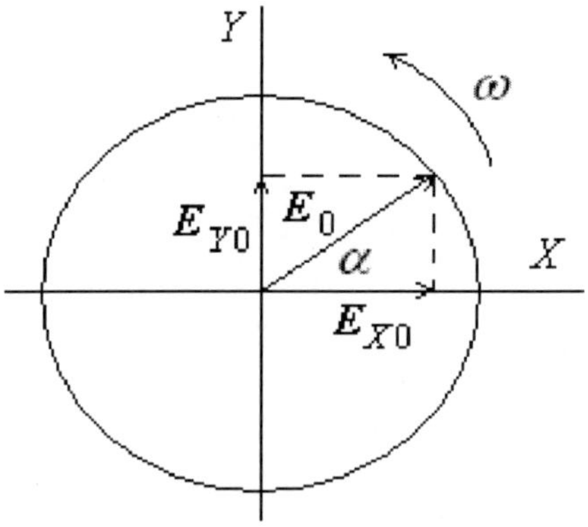

Figure 6.2.3.

Taking into account $\boldsymbol{E}(t)=\boldsymbol{E}_0 e^{-i\omega t}$, we shall find the peak value of a light vector of electromagnetic wave with circular polarization:

$$\boldsymbol{E}_0=\boldsymbol{E}_X+e^{i\varepsilon}\boldsymbol{E}_Y. \tag{6.2.4}$$

For light with only circular polarization, the module of amplitude at rotation keeps constant size $E_0=const$. But at the existence in structure of light components with linear polarization the peak values of light vector projections are not equal among themselves: and $\boldsymbol{E}_Y=\boldsymbol{E}_{Y0}\sin\alpha$, where angle and $\boldsymbol{E}_{X0}\neq\boldsymbol{E}_{Y0}$. Therefore, the most general formula for a light vector looks like:

$$\boldsymbol{E}_0=\boldsymbol{E}_{X0}\cos\alpha+e^{i\varepsilon}\boldsymbol{E}_{Y0}\sin\alpha. \tag{6.2.5}$$

The light intensity in this case equals:

$$I=c\varepsilon_0\frac{\boldsymbol{E}_0^2}{2}=c\varepsilon_0\frac{\boldsymbol{E}_0\boldsymbol{E}_0^*}{2}=,$$
$$=J_{XX}\cos^2\alpha+J_{YY}\sin^2\alpha+J_{XY}e^{-i\varepsilon}\cos\alpha\sin\alpha+J_{YX}e^{i\varepsilon}\sin\alpha\cos\alpha \tag{6.2.6}$$

where, for the convenience of the transformations, the conjugated light vector [43] is used.

Factor matrix at trigonometric functions (6.2.6):

$$J_{ij}=\begin{pmatrix} J_{XX} & J_{XY}e^{-i\varepsilon} \\ J_{YX}e^{i\varepsilon} & J_{YY} \end{pmatrix}=\frac{1}{2}\begin{pmatrix} E_{X0}E^*_{X0} & E_{X0}E^*_{Y0}e^{-i\varepsilon} \\ E_{Y0}E^*_{X0}e^{i\varepsilon} & E_{Y0}E^*_{Y0} \end{pmatrix} \quad (6.2.7)$$

is the tensor. The tensor components are not diagonal (6.2.7); complex and conjugated values $J_{ij}=J^*_{ji}$, i.e. the matrix (6.2.7) is a Hermitian matrix. They are not equal to zero; only at the existence of the light ray of elements of circular polarization.

The sum of the diagonal tensor components defines the module of intensity of light:

$$I=\mathrm{Sp}J_{ij}=J_{ii}=J_{XX}+J_{YY}.$$

The known principle of addition on repeating indexes is used.

To exclude the influence of intensity on the analysis of light polarization, we shall normalize the tensor (6.2.7) on light intensity *I*. We shall receive the so-called polarizing tensor:

$$\rho_{ij}=\frac{J_{ij}}{J_{ii}}=\begin{pmatrix} \rho_{XX} & \rho_{XY} \\ \rho_{YX} & \rho_{YY} \end{pmatrix}. \quad (6.2.8)$$

For a component of polarizing tensor the ratios are correct:

$$\rho_{XX}+\rho_{YY}=1 \text{ and } \rho_{XY}=\rho^*_{YX}. \quad (6.2.9)$$

1. We shall find the polarizing tensor for natural light in which there is circular polarization. The intensity of such light calculated under the formula (6.2.6) should not depend on angle α. It is possible only in this case: or alternatively: $\rho_{XX}=\rho_{YY}=\frac{1}{2}$. There are sizes or since there are no elements of circular polarization. Hence, for natural light the polarizing tensor equals:

$$\rho_{ij}=\frac{1}{2}\begin{pmatrix} 1 & 0 \\ 0 & 1 \end{pmatrix}. \quad (6.2.10)$$

Determinant of matrix $\left|\rho_{ij}\right|=\frac{1}{4}$.

2. We shall find the polarizing tensor for linearly polarized light. Elements of circular polarization are not present, i.e. $\rho_{XY}=\rho_{YX}=0$. If the direction of polarization coincides with the coordinates' axes in (6.2.7) is absent either component E_{Y0}, or a component E_{X0}. In this case the polarizing tensor looks like:

$$\rho_{ij} = \begin{pmatrix} 1 & 0 \\ 0 & 0 \end{pmatrix} \text{ or } \rho_{ij} = \begin{pmatrix} 0 & 0 \\ 0 & 1 \end{pmatrix}. \tag{6.2.11}$$

In both cases the determinant of matrix $|\rho_{ij}| = 0$.

The value determined by a condition [44] is referred to as the degree of linear polarization of light:

$$P = \sqrt{1 - 4|\rho_{ij}|}. \tag{6.2.12}$$

The degree of polarization varies from for natural light up to for linearly polarized light.

3. We shall find the polarizing tensor for light with circular polarization. In this case a phase of delay $\varepsilon = \frac{\pi}{2}$; hence, and $e^{-i\varepsilon} = \cos\varepsilon - i\sin\varepsilon = -i$. Besides taking into account the circular symmetry of the light vector, we have:

$$J_{XX} = J_{YY} = J_{XY} = J_{YX}. \tag{6.2.13}$$

The matrix (6.2.7) equals:

$$J_{ij} = \begin{pmatrix} J_{XX} & -iJ_{XY} \\ iJ_{YX} & J_{YY} \end{pmatrix}. \tag{6.2.14}$$

Hence, the polarizing tensor (if *Y*-components of light vector are delayed relative to the *X*-components) it equals:

$$\rho_{ij} = \frac{1}{2}\begin{pmatrix} 1 & -i \\ i & 1 \end{pmatrix}. \tag{6.2.15}$$

If *X*-components are delayed relative to the *Y*-components, the polarizing tensor equals:

$$\rho_{ji} = \frac{1}{2}\begin{pmatrix} 1 & i \\ -i & 1 \end{pmatrix}. \tag{6.2.16}$$

Tensors (6.2.15) and (6.2.16) correspond to the left and right circular polarizations.

Let's consider the problem of the degree of circular polarization.

Let's decompose the polarizing tensor on the symmetrical and antisymmetrical parts with the help of identity:

$$\rho_{ij} = \frac{1}{2}(\rho_{ij} + \rho_{ji}) + \frac{1}{2}(\rho_{ij} - \rho_{ji}). \tag{6.2.17}$$

For the component of the polarizing tensor of light polarized on a circle according to formulas (6.2.15) or (6.2.16) we have [43]:

$$\frac{1}{2}(\rho_{ij} + \rho_{ji}) = \frac{1}{2}\begin{pmatrix} 1 & 0 \\ 0 & 1 \end{pmatrix}, \tag{6.2.18}$$

i.e. it is natural light (6.2.10).

The second component (6.2.17) characterizes the circular polarization of light. For the component of circular polarization tensor we shall find:

$$\frac{1}{2}(\rho_{ij} - \rho_{ji}) = \frac{1}{2}\begin{pmatrix} 0 & -i \\ +i & 0 \end{pmatrix} = -\frac{i}{2}\begin{pmatrix} 0 & 1 \\ -1 & 0 \end{pmatrix}, \tag{6.2.19}$$

i.e. this component has only an imaginary value [44].

If light is only partly polarized on a circle instead of (6.2.19) it is necessary to write down:

$$\frac{1}{2}(\rho_{ij} - \rho_{ji}) = -\frac{i}{2}\begin{pmatrix} 0 & 1 \\ -1 & 0 \end{pmatrix} A, \tag{6.2.20}$$

where A is the degree of circular polarization.

The degree of circular polarization is in a range $-1 \le A \le 1$. Extreme limits correspond to the left and right circular polarization. At circular polarization is absent.

Generally, the degree of polarization of light is determined by two parameters: P - degree of linear polarization (6.2.12) and A - degree of circular polarization.

To find the degree of polarization which takes into account both linear and circular polarization, it is convenient to use Stokes' parameters which are dependent on the polarizing tensor components (6.2.8) under formulas:

$$\xi_1 = \rho_{XY} + \rho_{YX}\,;\ \xi_2 = i(\rho_{XY} - \rho_{YX});\ \xi_3 = \rho_{XX} - \rho_{YY}\,. \tag{6.2.21}$$

In this case, the polarizing tensor (6.2.8) receives:

$$\rho_{ij} = \frac{1}{2}\begin{pmatrix} 1+\xi_3 & \xi_1 - i\xi_2 \\ \xi_1 + i\xi_2 & 1-\xi_3 \end{pmatrix}. \tag{6.2.22}$$

The polarizing tensor determinant in Stokes' parameters equals:

$$\left|\rho_{ij}\right| = \frac{1}{4}\left(1 - \xi_1^2 - \xi_2^2 - \xi_3^2\right). \tag{6.2.23}$$

Hence, the degree of polarization, according to definition (6.2.12), equals:

$$P = \sqrt{\xi_1^2 + \xi_2^2 + \xi_3^2}\,. \tag{6.2.24}$$

The formula (6.2.24) is more general than (6.2.12), since it includes the component of a light vector with circular polarization.

Comparing (6.2.20) with (6.2.22), we shall find the top right component of the polarizing tensor (6.2.22). The Stokes' parameter $\xi_1 = 0$, since this parameter is real, and the top right component (6.2.20) is only imaginary. Hence:

$$\frac{1}{2}\left(\xi_1 - i\xi_2\right) = -\frac{i}{2}A. \tag{6.2.25}$$

Thus, $\xi_2 = A$. In the absence of circular polarization, i.e. $A = 0$, from (6.2.24) we receive the degree of linear polarization $P = \sqrt{\xi_1^2 + \xi_3^2}$.

Using (6.2.24), we find the module of the degree of polarization for light with any type of polarization:

$$\left|P_0\right| = \sqrt{P^2 + A^2}\,. \tag{6.2.26}$$

Hence, generally the degree of polarization of light has the complex character:

$$P_0 = P + iA\,, \tag{6.2.27}$$

where for the real part *P* the degree of linear polarization is characterized, and the imaginary part *A*- the degree of circular polarization.

6.3. The Quantum-Relativistic Form of Malus' Law

Malus' law (6.2.3) has been experimentally established by French physicist E.L. Malus who, being a supporter of Newton's corpuscular theory of light, tried to explain it with the help of moving light particles. Unfortunately, at that time this attempt was essentially unrealized due to the insufficient development of quantum representations about the nature of light.

The essence of the quantum approach to explain Malus' law consists in the following. At the interaction with the substance of the analyzer there is a certain probability of rotation of the quantum polarization plane on some angle. If the angular position of the quantum polarization plane after interaction with the substance of the analyzer coincides with the analyzer plane, the quantum will pass through the analyzer.

Thus, from the quantum point of view, Malus' law has a strongly pronounced probability character. Therefore, for finding the effective section of interaction of quantum with electrons of the analyzer, the quantum-relativistic analysis of Malus' law is quite adequate.

6.3.1. Interaction of Quantums of Electromagnetic Radiation with Electrons

The physical phenomenon reflected by Malus' law in quantum representation concerns the large group of phenomena of interaction of quantums of electromagnetic radiation with electrons.

Absorption by the substance of quantums of electromagnetic radiation basically results in the following processes: quantums can be absorbed without the occurrence of secondary electrons when all the quantum energy is transferred fonons - quantums of mechanical waves in a crystal lattice, and the impulse is transferred to all crystal lattice of substance [45], at an internal photoeffect, see paragraph 4.1, Compton's effect, see paragraph 4.2 and combinational scattering [4].

The latter two effects are accompanied by a secondary occurrence, i.e. reradiated quantums with changed frequency, in comparison with falling on the electron, and absorption of part energy of a falling photon.

In [18] it is shown that interaction of the polarized quantum of light with the non-polarized electron can lead to rotation of the polarizing plane of reradiated quantum, and to a change of frequency of the secondary quantum in comparison with the primary quantum. From the point of view of Malus' law, the phenomenon of rotation of the polarizing plane of reradiated quantum in comparison with the primary one enables this quantum to pass through the analyzer because the separation of quantum, as opposed to an electromagnetic wave, is considered impossible. A photoeffect and Compton's effect are basically inherent to the quantums' high energy. At the combinational scattering on coupled electrons in the field of seen light there is also a change of light frequency. At this form of interaction the absorption of a substance partly falling on photon energy is observed.

6.3.2. Physical Essence of Problem

At the passage of light through the analyzer there is an interaction of photons with coupled electrons in a crystal lattice of substance of the analyzer.

Interaction of electromagnetic radiation with coupled electrons of atoms or ions of a crystal lattice is complex and involves multiple physical processes. Therefore, we shall consider the elementary physical phenomenon – the interaction of electromagnetic radiation with free or poorly coupled initially motionless (for simplification) electrons at which the increase in the length of wave reradiated quantum can be observed. If the energy of falling radiation is great enough (for example, X-ray radiation), it is possible to neglect the coupling electrons in atoms and assume that electrons cooperate with radiation as free particles [21].

Taking into account that the quantum of light or elementary photon, i.e. an indivisible particle of electromagnetic field, direct division of the quantum is impossible. However, such division is present at Compton's effect (so we usually refer to the interaction of the quantum with motionless free electron), but it occurs by means of another elementary particle of substance – the electron.

Taking into account that the representations are far from complete, regarding both the nature of quantum and the nature of the electron, in modern physics it was necessary to enter a concept about "intermediate states" systems; a photon - electron during their interaction. These intermediate states are unusual enough. They are characterized, first, by an opportunity for negative electron energy, and second, by the infringement of the law of energy conservation.

Do these intermediate states have any physical sense or are they only mathematical concepts for calculation? An answer to this question has not been found until now. There is also no answer to the question of whether the mathematical receptions used are at the given physical preconditions, i.e. are our concepts of an intermediate state correct enough?

For a more detailed analysis of these questions we shall consider the transition of an initial state of a photon-electron system to its final state $F(\boldsymbol{k}, \boldsymbol{p}, \hbar\omega, E)$. In this case *S* (start) means an initial state of system *F* (finish) - its final state. Parameters of the system a photon - electron in an initial state: there is a vector of an impulse of a photon in an initial state, where $\hbar$ - Planck's reduced constant, - cyclic frequency of an initial photon, *c* - speed of light in vacuum, - a unit vector in a direction of radiation of an initial photon; - an impulse of an immovable electron in an initial state; - energy of a photon in an initial state; - energy of immovable electron in an initial state, - its rest mass.

The parameters of the system a photon - electron in a final state: there is a vector of an impulse of a photon in a final state, - cyclic frequency of a final photon, - a unit vector in a direction of radiation of a final photon; - an impulse of electron in a final state, - its speed, - its relativistic mass; - energy of a photon in a final state; - energy of the electron in a final state, *p* - the module of an impulse of electron in a final state. The transition of system a photon-electron from a state into a state is possible under the following schemes [18, 46]. At first the electron absorbs a photon and transits in an intermediate state with energy E_{I}. In this state there are no photons in the system. The electron in this case gets an impulse which is equal to an impulse of the absorbed photon. Then the electron emits a photon with an impulse $\boldsymbol{k}$, getting energy *E* and an impulse $\boldsymbol{p}$. The scheme of transition is shown with the help of Feynman's diagrams in figure 6.3.1. The straight lines designate movement of the electron, wavy - photons. The dotted line specifies that real movement of the electron in this case is absent.

The most unsatisfying element of this scheme is that it cannot be realized. First of all, it is experimentally established that there is no time interval between acts of absorption and the emission of a photon [26]. There is no instant equality energy or impulses of electron and the falling quantum.

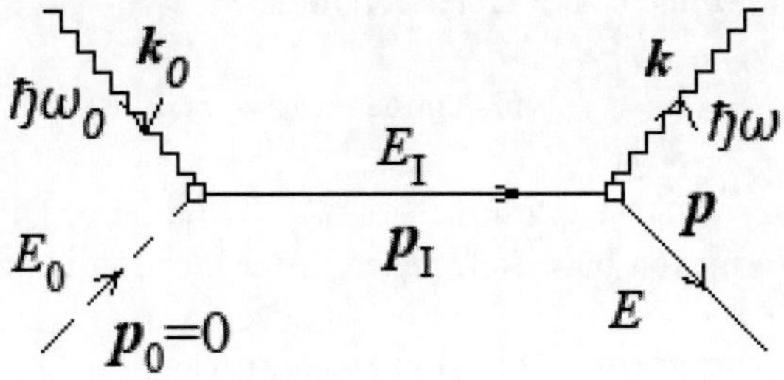

Figure 6.3.1.

However, there are other, more serious objections against the scheme in figure 6.3.1. The law of conservation of an impulse at absorption of a photon by electron in the scalar form looks like or $\frac{\hbar\omega_0}{c} = mV_{\mathrm{I}}$, where is a speed of electron in an intermediate state, m - its relativistic mass. The law of conservation of energy can be written down as:

$$\hbar\omega_0 = E_{\mathrm{I}} - E_0 = mc^2 - m_0c^2 = mc^2\left(1 - \frac{m_0}{m}\right) = mc^2\left(1 - \sqrt{1 - \frac{V_{\mathrm{I}}^2}{c^2}}\right).$$

Having divided the law of conservation of energy on the law of conservation of an impulse, we receive which, naturally, is impossible. Hence, one of two laws is not carried out. Usually, we can assume that the law of conservation of energy [4, 46] is not carried out. It results also in the non-execution of the law of conservation of energy and at radiation of a photon $\boldsymbol{k}$ since as a whole at the transition of the system a photon-electron from a state to a state $F(\boldsymbol{k}, \boldsymbol{p}, \hbar\omega, E)$, laws of conservation of energy and an impulse should be carried out. We shall note also that the energy of the electron in an intermediate state, formally, taking into account $p_{\mathrm{I}} = mV_{\mathrm{I}} = mc$, satisfies the condition $E_{\mathrm{I}} = \sqrt{p_{\mathrm{I}}^2c^2 + m_0^2c^4} \to \infty$, since at the mass $m \to \infty$.

With the purpose of overcoming the given difficulty, which is that free electron cannot absorb the quantum of light, in the photon - electron system is added, at the given stage of development of physics, actually the third substance; namely, a vacuum. We shall not pause on the theory of vacuum or its assumed structure; we shall note only that the surplus part of energy of electron is transferred for a while to a vacuum. Hence, the law of conservation of energy at the interaction of a photon and electron looks like $\hbar\omega_0 = E_{\mathrm{I}} + E_{vac} = \sqrt{p_{\mathrm{I}}^2c^2 + m_0^2c^4} + E_{vac}$. Then the energy of the vacuum returns in the photon - electron system, translating it into the state $F(\boldsymbol{k}, \boldsymbol{p}, \hbar\omega, E)$. If we are to accept the energy of the vacuum (so-called energy of electromagnetic vacuum), the impulse of the electron in an intermediate state (under condition of $k_0 = p_{\mathrm{I}}$) becomes equal to the imaginary expression $p_{\mathrm{I}} = \frac{2m_0 c}{\sqrt{3}} i$, which apparently is inadmissible.

Here, it is possible to rebuke the following. At the solution of Dirac's equation for the electron [47] there is information on the existence of the electron imaginary dipole moment $i\frac{\mu_0}{c}$, where is the Bohr magneton. Thus, the electromechanical ratio for the electron (by analogy with the magnetomechanical ratio) in the intermediate state appears in the real size $\frac{\mu_0 \sqrt{3}}{2m_0 c^2} = 0.98 \cdot 10^{-10}\, T^{-1}$. It specifies that the intermediate state can be quite real.

The used formula for vacuum energy formally follows from the analysis of the linear quantum oscillator [4]. The energy of such oscillator, which in quantum electrodynamics is identified with some sets of possible energies of quantums of electromagnetic field, equals $E = \left(n + \frac{1}{2} \right) \hbar\omega_0$, where n is a serial number of oscillator energy level. Each energy level is identified with the possible existence of some photon; i.e. it can be connected to a real photon, but can also be empty. At the zero level the energy appears not equal to zero, but any photons in space are obviously not present. This energy is identified with vacuum energy.

In figure 6.3.2 the set of energy levels of the linear quantum oscillator or the possible set of photons energies is shown. Between energy levels there are identical intervals with energy $\varepsilon = \hbar\omega_0$. Except for the energy levels with positive energy, according to Dirac, energy levels with negative energies [48] are entered.

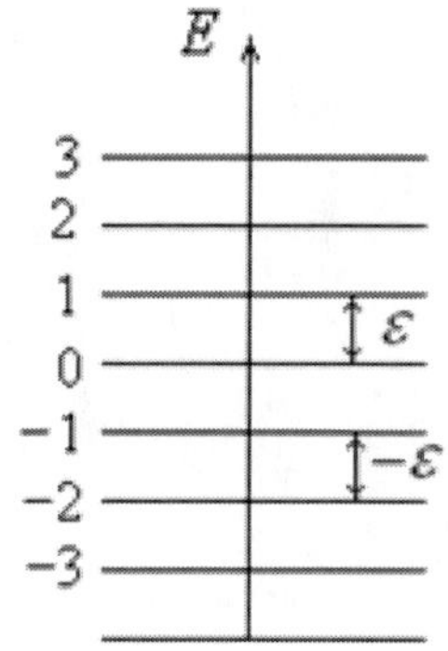

Figure 6.3.2.

We assume that there are no real particles (bosons or fermions) at examined energy levels (positive and negative) or in considered space, i.e. they empty - vacuum. Vacuum energy can be found by the direct addition of possible energies of the linear oscillator, figure 6.3.2:

$$E_{vac} = \varepsilon - \varepsilon + \varepsilon - \varepsilon + \varepsilon - ... = \varepsilon(1 - 1 + 1 - 1 + 1 - ...).$$

Strictly speaking, the occurring sign-variable infinite series is divergent. This series is divergent in the sense of absence of the sequence limit of the partial sums. However, it does not mean that it has no sum. From the end of the 19th century to the beginning of the 20th century the theory of summation of infinite sign-variable divergent series was created. For example, in [49] the given series is summarized to size by the Abel - Poisson method.

But this series is possible to summarize very simply, having designated $1 - 1 + 1 - 1 + 1 - ... = S$. Transferring the first unit in the right part and changing a sign on both parties of equality, we find $S = -S + 1$. Hence, $S = \frac{1}{2}$. Thus, it is possible to find the formula for calculation of vacuum energy $E_{vac} = \frac{1}{2}\varepsilon$. Further, using the distance between energy levels $\varepsilon = \hbar\omega_0$, we find the vacuum energy $E_{vac} = \frac{1}{2}\hbar\omega_0$.

The resulting method of reception of vacuum energy evidently shows that this energy is the direct consequence of quantum representations about the structure of substance and the electromagnetic field. If the quantum properties - for example, the energy of electromagnetic field - were absent ($\varepsilon = 0$) from the following analysis, that energy of the vacuum would be zero.

It is possible to assume the energy of the vacuum where in this case *m* is a whole number which is more than unit – quantity of linear quantum oscillators. In this case electron energy in the intermediate state according to the formula to become zero (at $m = 2$) or negative (at $m > 2$), since $E_{\text{I}} = \left(1 - \frac{m}{2}\right)\hbar\omega_0$.

Usually, negative electron energy in the intermediate state is formally connected with an opportunity of a minus sign before a root in the expression for energy [46].

Something should be said about the so-called renormalization problem, as seen by one of the founders of the given theory [50]. Frequently, renormalization is represented as the final solution of all unclear positions of quantum electrodynamics. Actually, it is not so. For example, in quantum electrodynamics there is the problem of electron mass. A moving electron creates its own electromagnetic field around itself.

This field cooperates with the electron. The process is referred to as a reaction of the field (radiation). The mass caused by the reaction of the field is referred to as electromagnetic electron mass.

According to the results of the calculations, the electromagnetic electron mass has an indefinitely great size since it is proportional to the energy of its own field; hence, $m_e \sim \frac{1}{r}$, where *r* is the distance from the electron's center. It contradicts experiments. Renormalization will consist in the procedure of the replacement of unacceptable calculated sizes by experimental values. This procedure on expression [50] has phenomenological character; therefore, it cannot explain the physical problems underlying quantum electrodynamics.

6.3.3. Malus' Law. The Quantum Approach

We shall begin quantum deduction of Malus' law with the analysis of the interaction of photons and electron in the analyzer. Thus, we use the approximation of free electrons [21]. For the deduction we shall not pay attention to physical nonsense, which was discussed earlier. All transformations we shall carry out in the CGS system.

Research of the processes of electron transition in intermediate state is rather simple because of the linearity of Hamilton's operator concerning value [4]. Using Coulomb's calibrations of potentials, we have:

$$\hat{H} = c\alpha\left(\hat{\boldsymbol{p}} - \frac{e}{c}\hat{\boldsymbol{A}}\right) + m_0 c^2 \beta = c\boldsymbol{\alpha}\hat{\boldsymbol{p}} - e\boldsymbol{\alpha}\hat{\boldsymbol{A}} + e\phi + m_0 c^2 \beta, \tag{6.3.1}$$

where $\boldsymbol{A}$ is a vector-potential of a photon, ϕ - its scalar potential, e - a charge of electron, - a matrix vector (a vector, whose coordinate projections are square matrixes of 4th order, with components 0, ±1, ± i), β - a square matrix of 4th order with components 0, ±1. All factors of matrix vector $\boldsymbol{\alpha}$ are four-vectors, of which the fourth component is imaginary [4]. The energy of interaction with the photon-electron characterizes only the second addend (6.3.1).

In (6.3.1) it is supposed that the vector-potential of a photon is the periodic function of coordinates $\boldsymbol{A} = \sqrt{\frac{2\pi\hbar c^2}{\omega_0 V}}\boldsymbol{e}_0 e^{-\frac{i}{\hbar}\boldsymbol{k}_0\boldsymbol{r}}$, there is unit vector of polarization of an initial photon, V - some normalizing volume. Therefore, it is possible to write down:

$$\hat{H}_{\text{int}} = -e\left(\boldsymbol{\alpha}\hat{\boldsymbol{A}}\right) = -e\sqrt{\frac{4\pi\hbar c^2}{\omega_0 V}}\left(\boldsymbol{\alpha}\boldsymbol{e}_0\right)e^{-\frac{i}{\hbar}\boldsymbol{k}_0\boldsymbol{r}}. \tag{6.3.2}$$

The matrix element of transition of the electron states looks like:

$$\begin{aligned}
\left(H_{\text{int}}\right)_{SI} &= \int \varphi_0^* \hat{H}_{\text{int}} \varphi_{\text{I}} dV = -e\sqrt{\frac{2\pi\hbar c^2}{\omega_0 V}} \int \varphi_0^* \left(\boldsymbol{\alpha}\boldsymbol{e}_0\right) e^{-\frac{i}{\hbar}\boldsymbol{k}_0\boldsymbol{r}} \varphi_{\text{I}} dV = \\
&= -e\sqrt{\frac{2\pi\hbar c^2}{\omega_0 V}} \int u_0^* e^{-i\frac{\boldsymbol{p}_0\boldsymbol{r}}{\hbar}} \left(\boldsymbol{\alpha}\boldsymbol{e}_0\right) e^{-\frac{i}{\hbar}\boldsymbol{k}_0\boldsymbol{r}} u_{\text{I}} e^{i\frac{\boldsymbol{p}_1\boldsymbol{r}}{\hbar}} dV = \\
&= -e\sqrt{\frac{2\pi\hbar c^2}{\omega_0 V}} \left[u_0^* \left(\boldsymbol{\alpha}\boldsymbol{e}_0\right) u_{\text{I}}\right] \int e^{\frac{i}{\hbar}\left(\boldsymbol{p}_1 - \boldsymbol{p}_0 - \boldsymbol{k}_0\right)\boldsymbol{r}} dV
\end{aligned}, \tag{6.3.3}$$

where is the conjugate amplitude of wave function in the initial state of the electron, $\boldsymbol{r}$ - its radius - vector. After the absorption of a photon, we believe that the position of the electron not changed, is the amplitude of wave function in intermediate state. Taking into account the law of conservation of impulse $\boldsymbol{p}_{\mathrm{I}} - \boldsymbol{k}_0 - \boldsymbol{p}_0 = 0$, we have $\int e^{\frac{i}{\hbar}(\boldsymbol{p}_{\mathrm{I}} - \boldsymbol{p}_0 - \boldsymbol{k}_0)\boldsymbol{r}} dV = V$. Believing normalizing volume $V = 1$, we shall find a matrix element of transition $S \rightarrow \mathrm{I}$:

$$(H_{\text{int}})_{S\mathrm{I}} = -e\sqrt{\frac{2\pi\hbar^2 c}{k_0}}\left[\overset{*}{u_0}(\boldsymbol{a}\boldsymbol{e}_0)u_{\mathrm{I}}\right]. \tag{6.3.4}$$

The received expression corresponds, for example, [46].

Let's consider the process of radiation of a photon $\boldsymbol{k}$ in the scheme on figure 6.3.1, i.e. in process $\mathrm{I} \rightarrow F$. Similar to the previous, the transformations allow us to receive a matrix element of transition $\mathrm{I} \rightarrow F$:

$$(H_{\text{int}})_{\mathrm{I}F} = -e\sqrt{\frac{2\pi\hbar^2 c}{k}}\left[\overset{*}{u_{\mathrm{I}}}(\boldsymbol{a}\boldsymbol{e})u\right], \tag{6.3.5}$$

where is the conjugate amplitude of wave function of the electron in the intermediate state, u - amplitude of wave function of the electron in a final condition, - a unit vector of polarization of a final photon. For the full description of the transition of a photon-electron system, it is necessary to consider one more formal variant of the process which can be represented in the diagram of Feynman shown in figure 6.3.3.

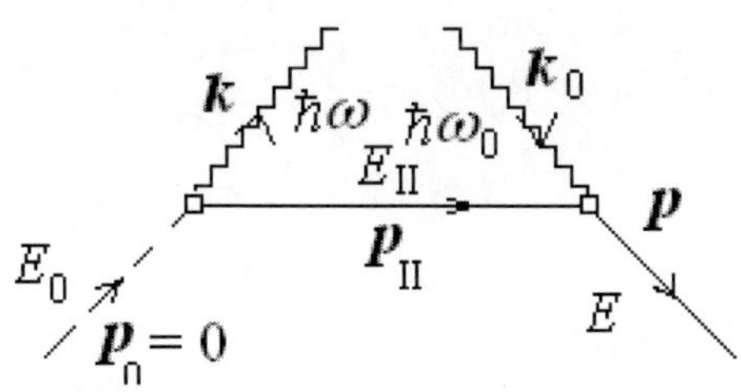

Figure 6.3.3.

According to the considered scheme of transition, figure 6.3.3, motionless ($\boldsymbol{p}_0 = 0$) electron having energy E_0, again radiates the quantum an impulse $\boldsymbol{k}$ and energy and passes in the intermediate state II. Then the electron absorbs the initial quantum with impulse and energy $\hbar\omega_0$, getting an impulse $\boldsymbol{p}$ and energy E. The matrix element of transition with radiation of a photon looks like:

$$(H_{\text{int}})_{S\mathrm{II}} = -e\sqrt{\frac{2\pi\hbar^2 c}{k}}\left[\overset{*}{u_0}(\boldsymbol{a}\boldsymbol{e})u_{\mathrm{II}}\right], \tag{6.3.6}$$

and the matrix element of transition with absorption of a photon looks like:

$$\left(H_{\text{int}}\right)_{\text{II}F} = -e\sqrt{\frac{2\pi\hbar^2 c}{k_0}}\left[\overset{*}{u}_{\text{II}}\left(\boldsymbol{\alpha e}_0\right)u\right], \tag{6.3.7}$$

where and are the conjugate and primary amplitudes of wave function of the electron in the intermediate state. The compound matrix element determining full probability of transition looks like:

$$\left(H_{\text{int}}\right)_{SF} = \frac{\left(H_{\text{int}}\right)_{S\text{I}}\left(H_{\text{int}}\right)_{\text{I}F}}{E_0 - E_{\text{I}} + \hbar\omega_0} + \frac{\left(H_{\text{int}}\right)_{S\text{II}}\left(H_{\text{int}}\right)_{\text{II}F}}{E_0 - E_{\text{II}} - \hbar\omega}. \tag{6.3.8}$$

Let's substitute in (6.3.8) formulas (6.3.4) - (6.3.7):

$$\left(H_{\text{int}}\right)_{SF} = e^2\frac{2\pi\hbar^2 c}{\sqrt{k_0 k}}\left\{\begin{aligned}&\frac{\left[\overset{*}{u}_0\left(\boldsymbol{\alpha e}_0\right)u_{\text{I}}\right]\left[\overset{*}{u}_{\text{I}}\left(\boldsymbol{\alpha e}\right)u\right]}{E_0 - E_{\text{I}} + \hbar\omega_0} + \\ &+\frac{\left[\overset{*}{u}_0\left(\boldsymbol{\alpha e}\right)u_{\text{II}}\right]\left[\overset{*}{u}_{\text{II}}\left(\boldsymbol{\alpha e}_0\right)u\right]}{E_0 - E_{\text{II}} - \hbar\omega}\end{aligned}\right\}. \tag{6.3.9}$$

The probability of transition is in unit of time proportional to a square of a corresponding matrix element [4]:

$$dW = \frac{2\pi}{\hbar}\left|\left(H_{\text{int}}\right)_{SF}\right|^2 \rho_F(\omega)d\Omega, \tag{6.3.10}$$

where is the number of the final state of the photon-electron system, contained in a unit interval of energy of a final state [46], - a solid angle describing a direction of a start of reradiated photon. The number of final states of reradiated photon contained in the interval of photon energy corresponding to an impulse interval is equal $\rho(\omega)dk$, where - number of states of the photons contained in a unit interval of their energy [4]. Naturally, the number of states of photons will define the number of states of all systems:

$$\rho_F(\omega)dE_F = c\rho(\omega)dk, \tag{6.3.11}$$

where is an energy interval of a final state of the system, corresponding to impulse interval of photons dk, and final energy of system photon - electron $E_F = ck + E = ck + \sqrt{p^2c^2 + m_0^2c^4}$.

Hence:

$$\rho_F(\omega) = c\rho(\omega)\frac{\partial k}{\partial E_F}. \tag{6.3.12}$$

Substituting (6.3.12) in (6.3.10), we shall find the probability of transition in unit of time:

$$dW = \frac{2\pi}{\hbar}\left|(H_{\text{int}})_{SF}\right|^2 c\rho(\omega)\frac{\partial k}{\partial E_F}d\Omega. \tag{6.3.13}$$

For finding the differential effective section of process, we shall divide the probability of transition in unit of time into numerical density of a flux of falling quantums [4].

By determination of the differential effective section of process $d\sigma = \frac{dP}{I_0}$, see (3.1.4), where is a flux of energy of scattering radiation in a solid angle $d\Omega$, - intensity of falling radiation. We shall divide the numerator and denominator into energy of quantum $\hbar\omega_0$. Taking into account and $j_0 = \frac{I_0}{\hbar\omega_0}$, we shall find $d\sigma = \frac{dW}{j_0}$. The given formula represents the determination of differential effective section of process at the quantum approach. One quantum is absorbed; therefore, we assume that in normalizing volume V there is only one quantum. The intensity of falling radiation in this case equals $I_0 = \frac{\hbar\omega_0}{V}c$. Therefore, the numerical density of a flux of photons under condition $V = 1$, equals $\frac{I_0}{\hbar\omega_0} = \frac{c}{V} = c$.

Hence, the differential effective section of process $S \to F$:

$$d\sigma = \frac{dW}{j_0} = \frac{2\pi}{\hbar}\left|(H_{\text{int}})_{SF}\right|^2 \rho(\omega)\frac{\partial k}{\partial E_F}d\Omega, \tag{6.3.14}$$

Further substituting (6.3.9) in (6.3.14), with the account $\rho(\omega) = \frac{k^2}{(2\pi\hbar)^3 c}$, we shall find:

$$d\sigma = \frac{1}{2}e^4\frac{k}{k_0}\frac{c\partial k}{\partial E_F}\left\{\begin{array}{l}\dfrac{\left[u_0^*(\boldsymbol{\alpha e}_0)u_{\text{I}}\right]\left[u_{\text{I}}^*(\boldsymbol{\alpha e})u\right]}{E_0 - E_{\text{I}} + \hbar\omega_0} + \\ + \dfrac{\left[u_0^*(\boldsymbol{\alpha e})u_{\text{II}}\right]\left[u_{\text{II}}^*(\boldsymbol{\alpha e}_0)u\right]}{E_0 - E_{\text{II}} - \hbar\omega}\end{array}\right\}^2 d\Omega. \tag{6.3.15}$$

Factor in (6.3.15) is the consequence of averaging the differential effective section of interaction of the photon-electron on spin states of the electron in its initial state. It is similar, for example, to the averaging of differential effective section of scattering of the non-polarized wave on a vector of polarization of electromagnetic wave at the deduction of Thomson's formula (3.2.8). Let's find matrix elements in the formula (6.3.15). We shall consider the first addend in brackets. We shall multiply the numerator and denominator in it on $E_0 + E_\text{I} + \hbar\omega_0$. As a result, we shall obtain:

$$\frac{\left[u_0^*(\boldsymbol{\alpha e}_0)u_\text{I}\right]\left[u_\text{I}^*(\boldsymbol{\alpha e})u\right]}{(E_0 - E_\text{I} + \hbar\omega_0)} = \frac{\left[u_0^*(\boldsymbol{\alpha e}_0)u_\text{I}\right]\left[u_\text{I}^*(\boldsymbol{\alpha e})u\right](E_0 + E_\text{I} + \hbar\omega_0)}{(E_0 - E_\text{I} + \hbar\omega_0)(E_0 + E_\text{I} + \hbar\omega_0)} = \frac{\left[u_0^*[(\boldsymbol{\alpha e}_0)(\boldsymbol{\alpha e})]u\right](E_0 + \hbar\omega_0 + E_\text{I})}{2E_0\hbar\omega_0}. \quad (6.3.16)$$

In the numerator the known formula is used, and also $p_\text{I} = k_0 = \dfrac{\hbar\omega_0}{c}$.

Similarly, the multiplication of the numerator and denominator on with the account $-p_\text{II} = k = \dfrac{\hbar\omega}{c}$, will transform the second addend (6.3.15). Hence, the formula (6.3.15) takes the form:

$$d\sigma = \frac{e^4}{8c^2E_0^2}\frac{k}{k_0}\frac{c\partial k}{\partial E_F}\left\{\frac{\left[u_0^*[(\boldsymbol{\alpha e}_0)(\boldsymbol{\alpha e})]u\right](E_0 + \hbar\omega_0 + E_\text{I})}{k_0} - \frac{\left[u_0^*[(\boldsymbol{\alpha e})(\boldsymbol{\alpha e}_0)]u\right](E_0 - \hbar\omega + E_\text{II})}{k}\right\}^2 d\Omega. \quad (6.3.17)$$

In intermediate states with energies and the electron is a free particle which the electromagnetic field does not influence as the photon is already absorbed or radiated by the electron; therefore, the amplitude of its wave function satisfies Dirac's equation for free particle [4]:

$$E_\text{I}u_\text{I} = \hat{H}_\text{I}u_\text{I} \text{ and } E_\text{II}u_\text{II} = \hat{H}_\text{II}u_\text{II}, \quad (6.3.18)$$

where Hamilton's operators follow the formula (6.3.1) at since in intermediate states there are no photons:

$$\hat{H}_\text{I} = c\boldsymbol{\alpha}\hat{\boldsymbol{p}}_\text{I} + m_0c^2\beta = c\boldsymbol{\alpha}\hat{\boldsymbol{k}}_0 + E_0\beta, \quad (6.3.19)$$

$$\hat{H}_{\text{II}} = c\boldsymbol{\alpha}\hat{\boldsymbol{p}}_{\text{II}} + m_0 c^2 \beta = -c\boldsymbol{\alpha}\hat{\boldsymbol{k}} + E_0 \beta, \tag{6.3.20}$$

where it is taken into account $\boldsymbol{p}_{\text{I}} = \boldsymbol{k}_0$, figure 6.3.1, and $\boldsymbol{p}_{\text{II}} = -\boldsymbol{k}$, figure 6.3.3. Hence, it is possible to assume and $E_{\text{II}} + E_0 = -c\boldsymbol{\alpha k}$, where is the initial level of reading of energy equal to the initial energy of the motionless electron. Components of a matrix $\beta = \pm 1$. $\beta = -1$ is used, since initial and intermediate energy in additional factors and are summarized. Substituting these values in (6.3.17), we shall find:

$$d\sigma = \frac{e^4}{8c^2 E_0{}^2} \frac{k}{k_0} \frac{c\partial k}{\partial E_F} \left\{ \begin{array}{l} \dfrac{\left[u_0^*\left[(\boldsymbol{\alpha e}_0)(\boldsymbol{\alpha e})\right]u\right](\hbar\omega_0 + c(\boldsymbol{\alpha k}_0))}{k_0} - \\ - \dfrac{\left[u_0^*\left[(\boldsymbol{\alpha e})(\boldsymbol{\alpha e}_0)\right]u\right](-\hbar\omega - c(\boldsymbol{\alpha k}))}{k} \end{array} \right\}^2 d\Omega =$$

$$= e^4 \frac{1}{2} \frac{k}{k_0} \frac{c\partial k}{\partial E_F} \left\{ \frac{1}{2E_0} u_0^* \left[\begin{array}{l} 2(\boldsymbol{e}_0\boldsymbol{e}) - \dfrac{(\boldsymbol{\alpha e}_0)(\boldsymbol{\alpha k}_0)(\boldsymbol{\alpha e})}{k_0} - \\ - \dfrac{(\boldsymbol{\alpha e})(\boldsymbol{\alpha k})(\boldsymbol{\alpha e}_0)}{k} \end{array} \right] u \right\}^2 d\Omega =$$

$$= e^4 \frac{1}{2} \frac{k}{k_0} \frac{c\partial k}{\partial E_F} \sum \left| u_0^* Q u \right|^2 d\Omega, \tag{6.3.21}$$

where it is designated:

$$Q = \frac{1}{2E_0} \left[2(\boldsymbol{e}_0\boldsymbol{e}) - \frac{(\boldsymbol{\alpha e}_0)(\boldsymbol{\alpha k}_0)(\boldsymbol{\alpha e})}{k_0} - \frac{(\boldsymbol{\alpha e})(\boldsymbol{\alpha k})(\boldsymbol{\alpha e}_0)}{k} \right]. \tag{6.3.22}$$

Known rules of work with a matrix vector [21, 46] are used. The property of a matrix vector is used. Besides the following properties of action of a matrix vector on mutually perpendicular vectors $(\boldsymbol{\alpha e}_0)(\boldsymbol{\alpha k}_0) = -(\boldsymbol{\alpha k}_0)(\boldsymbol{\alpha e}_0)$, $(\boldsymbol{\alpha e})(\boldsymbol{\alpha k}) = -(\boldsymbol{\alpha k})(\boldsymbol{\alpha e})$, and also an opportunity for the cyclic rearrangement of factors with matrix vector is used. In the further transformations the ratio $(\boldsymbol{\alpha e}_0)^2 = (\boldsymbol{\alpha e})^2 = 1$, $(\boldsymbol{\alpha k})^2 = k^2$, will also be necessary.

The summation in (6.3.21) is carried out on two spin, real states of the electron with positive energy. For distribution of summation on all four theoretically possible states of the electron (2 spin and 2 signs on energy) are entered into the projective operator (at $\beta = 1$) [4]:

$$\hat{R} = \frac{\hat{H} + |E|}{2|E|} = \frac{c\boldsymbol{\alpha p} + \beta m_0 c^2 + |E|}{2|E|} = \\ \frac{c\boldsymbol{\alpha p} + E_0 + |E|}{2|E|} = \frac{E + |E|}{2|E|} \begin{cases} 1, & E > 0 \\ 0, & E < 0 \end{cases} , \tag{6.3.23}$$

where $E = E_0 + c\boldsymbol{\alpha p}$.

At the use of the projective operator a state with negative energy does not bring the contribution to a total sum. Hence, the expression (6.3.21) can be written down as:

$$d\sigma = e^4 \frac{1}{2} \frac{k}{k_0} \frac{c\partial k}{\partial E_F} \sum \left| u_0^* Q u \right|^2 d\Omega = \\ = e^4 \frac{1}{2} \frac{k}{k_0} \frac{c\partial k}{\partial E_F} \sum \left\{ u_0^* Q \hat{R} u \right\}\left\{ u^* Q^+ \hat{R}_0 u_0 \right\} d\Omega = \\ = e^4 \frac{1}{2} \frac{k}{k_0} \frac{c\partial k}{\partial E_F} \sum \left\{ u_0^* Q \hat{R} Q^+ \hat{R}_0 u_0 \right\} d\Omega = \\ = e^4 \frac{1}{2} \frac{k}{k_0} \frac{c\partial k}{\partial E_F} \sum \left\{ u_0^* Q \hat{R} Q^+ u_0 \right\} d\Omega , \tag{6.3.24}$$

where is the conjugate operator as a result of the replacement of sequence factors of the matrix operator in (6.3.22) on the return and change of a sign before addends with matrix vector $\boldsymbol{\alpha}$ [18]. The projective operator because the initial impulse of the electron is equal to zero $\boldsymbol{p}_0 = 0$. The projective operator of the free electron, according to a form of Hamilton's operator (6.3.1) in Dirac's equation (at and $\beta = 1$) in a final state, taking into account $\boldsymbol{p} = \boldsymbol{k}_0 - \boldsymbol{k}$, equals:

$$\hat{R} = \frac{|-(E_0 + c\boldsymbol{\alpha p})| + E_0 + c\boldsymbol{\alpha p}}{2|E|} = \frac{2E_0 + c(k_0 - k) + c\boldsymbol{\alpha}(\boldsymbol{k}_0 - \boldsymbol{k})}{2|E|} . \tag{6.3.25}$$

Substituting (6.3.22) and (6.3.25) in (6.3.24), we shall find:

$$d\sigma = e^4 \frac{1}{16 E_0{}^2 E} \frac{k}{k_0} \frac{c\partial k}{\partial E_F} \sum \left\{ \begin{gathered} u_0^* \left[\begin{gathered} 2(\boldsymbol{e}_0 \boldsymbol{e}) - \frac{(\boldsymbol{\alpha e}_0)(\boldsymbol{\alpha k}_0)(\boldsymbol{\alpha e})}{k_0} - \\ - \frac{(\boldsymbol{\alpha e})(\boldsymbol{\alpha k})(\boldsymbol{\alpha e}_0)}{k} \end{gathered} \right] \cdot \\ \cdot \left[2E_0 + c(k_0 - k) + c\boldsymbol{\alpha}(\boldsymbol{k}_0 - \boldsymbol{k}) \right] \cdot \\ \cdot \left[\begin{gathered} 2(\boldsymbol{e}_0 \boldsymbol{e}) + \frac{(\boldsymbol{\alpha e})(\boldsymbol{\alpha k}_0)(\boldsymbol{\alpha e}_0)}{k_0} + \\ + \frac{(\boldsymbol{\alpha e}_0)(\boldsymbol{\alpha k})(\boldsymbol{\alpha e})}{k} \end{gathered} \right] u_0 \end{gathered} \right\} d\Omega . \tag{6.3.26}$$

For the calculation of the expression under the symbol of the sum we shall take advantage of a rule spur $\sum \overset{*}{u}_0 \hat{L} u_0 = \mathrm{Sp}L$; for example [4], where is some matrix operator.

After the multiplication of brackets, we use the following ratio from properties of a matrix vector: an opportunity of reverse rearrangement of factors with $\boldsymbol{\alpha}$ and $(\boldsymbol{\alpha e})(\boldsymbol{\alpha e}_0) = (\boldsymbol{\alpha e}_0)(\boldsymbol{\alpha e}) = (\boldsymbol{e}_0\boldsymbol{e})$. As a result we have:

$$-\frac{(\boldsymbol{\alpha e}_0)(\boldsymbol{\alpha k}_0)(\boldsymbol{\alpha e})}{k_0} J \frac{(\boldsymbol{\alpha e})(\boldsymbol{\alpha k}_0)(\boldsymbol{\alpha e}_0)}{k_0} - \\ -\frac{(\boldsymbol{\alpha e})(\boldsymbol{\alpha k})(\boldsymbol{\alpha e}_0)}{k} J \frac{(\boldsymbol{\alpha e}_0)(\boldsymbol{\alpha k})(\boldsymbol{\alpha e})}{k} = -2(\boldsymbol{e}_0\boldsymbol{e})^2 J,$$

where $J = 2E_0 + c(k_0 - k)$, and also:

$$\frac{(\boldsymbol{\alpha e}_0)(\boldsymbol{\alpha k}_0)(\boldsymbol{\alpha e})}{k_0} J \frac{(\boldsymbol{\alpha e}_0)(\boldsymbol{\alpha k})(\boldsymbol{\alpha e})}{k} + \frac{(\boldsymbol{\alpha e})(\boldsymbol{\alpha k})(\boldsymbol{\alpha e}_0)}{k} J \frac{(\boldsymbol{\alpha e})(\boldsymbol{\alpha k}_0)(\boldsymbol{\alpha e}_0)}{k_0} = 0 \cdot$$

Carrying out simple, but bulky transformations, we use rules of work with spurs [3]: if $\boldsymbol{b}$ and $\boldsymbol{c}$ are some vectors, and is a matrix vector, $\frac{1}{4}\mathrm{Sp}(1) = 1$, $\mathrm{Sp}(\boldsymbol{b}+\boldsymbol{c}) = \mathrm{Sp}(\boldsymbol{b}) + \mathrm{Sp}(\boldsymbol{c})$, $\mathrm{Sp}(\boldsymbol{bc}) = \mathrm{Sp}(\boldsymbol{cb})$, $\mathrm{Sp}(\gamma\boldsymbol{b}) = \gamma\mathrm{Sp}(\boldsymbol{b})$, where γ is an algebraic function, $\frac{1}{4}\mathrm{Sp}(\boldsymbol{\alpha b})(\boldsymbol{\alpha c}) = \boldsymbol{bc}$. Besides the spur of addend with an odd number of factors containing a matrix vector it is equal to zero. Thus, (6.3.26) will be transformed:

$$d\sigma = e^4 \frac{1}{16E_0{}^2 E} \frac{k}{k_0} \frac{c\partial k}{\partial E_F} \mathrm{Sp}\left[\begin{array}{l} 2(\boldsymbol{e}_0\boldsymbol{e})^2 J + \\ + c\boldsymbol{\alpha}(\boldsymbol{k}_0 - \boldsymbol{k})\left[\begin{array}{l} -2(\boldsymbol{e}_0\boldsymbol{e})\frac{(\boldsymbol{\alpha e}_0)(\boldsymbol{\alpha k}_0)(\boldsymbol{\alpha e})}{k_0} - \\ -2(\boldsymbol{e}_0\boldsymbol{e})\frac{(\boldsymbol{\alpha e})(\boldsymbol{\alpha k})(\boldsymbol{\alpha e}_0)}{k} + \\ +\frac{(\boldsymbol{\alpha e})(\boldsymbol{\alpha k}_0)(\boldsymbol{\alpha e}_0)}{k_0} 2(\boldsymbol{e}_0\boldsymbol{e}) + \\ +\frac{(\boldsymbol{\alpha e}_0)(\boldsymbol{\alpha k})(\boldsymbol{\alpha e})}{k} 2(\boldsymbol{e}_0\boldsymbol{e}) \end{array}\right] \end{array}\right] d\Omega$$

Taking into account that it is possible to reverse the order of factors with a matrix vector under a symbol of spur, or to rearrange them cyclically, and also the linearity of a matrix vector $\boldsymbol{\alpha}(\boldsymbol{k}_0 - \boldsymbol{k}) = (\boldsymbol{\alpha k}_0) - (\boldsymbol{\alpha k})$, leads us to the multiplication:

$$d\sigma = e^4 \frac{1}{16E_0{}^2 E}\frac{k}{k_0}\frac{c\partial k}{\partial E_F}\mathrm{Sp}\left[\begin{array}{l} 2(\boldsymbol{e}_0\boldsymbol{e})^2 J + \\ +c\left[\begin{array}{l} -2(\boldsymbol{e}_0\boldsymbol{e})\dfrac{(\boldsymbol{\alpha k}_0)(\boldsymbol{\alpha e}_0)(\boldsymbol{\alpha k}_0)(\boldsymbol{\alpha e})}{k_0} \\ -2(\boldsymbol{e}_0\boldsymbol{e})\dfrac{(\boldsymbol{\alpha k}_0)(\boldsymbol{\alpha e})(\boldsymbol{\alpha k})(\boldsymbol{\alpha e}_0)}{k}+ \\ +\dfrac{(\boldsymbol{\alpha k}_0)(\boldsymbol{\alpha e})(\boldsymbol{\alpha k}_0)(\boldsymbol{\alpha e}_0)}{k_0}2(\boldsymbol{e}_0\boldsymbol{e})+ \\ +\dfrac{(\boldsymbol{\alpha k}_0)(\boldsymbol{\alpha e}_0)(\boldsymbol{\alpha k})(\boldsymbol{\alpha e})}{k}2(\boldsymbol{e}_0\boldsymbol{e})+ \\ +2(\boldsymbol{e}_0\boldsymbol{e})\dfrac{(\boldsymbol{\alpha k})(\boldsymbol{\alpha e}_0)(\boldsymbol{\alpha k}_0)(\boldsymbol{\alpha e})}{k_0}+ \\ +2(\boldsymbol{e}_0\boldsymbol{e})\dfrac{(\boldsymbol{\alpha k})(\boldsymbol{\alpha e})(\boldsymbol{\alpha k})(\boldsymbol{\alpha e}_0)}{k}- \\ -\dfrac{(\boldsymbol{\alpha k})(\boldsymbol{\alpha e})(\boldsymbol{\alpha k}_0)(\boldsymbol{\alpha e}_0)}{k_0}2(\boldsymbol{e}_0\boldsymbol{e})- \\ -\dfrac{(\boldsymbol{\alpha k})(\boldsymbol{\alpha e}_0)(\boldsymbol{\alpha k})(\boldsymbol{\alpha e})}{k}2(\boldsymbol{e}_0\boldsymbol{e}) \end{array}\right]\end{array}\right]d\Omega .$$

Then, allocating to each one two addends in internal square brackets a scalar factor and taking it out of these brackets, we find:

$$d\sigma = e^4 \frac{1}{16E_0{}^2 E}\frac{k}{k_0}\frac{c\partial k}{\partial E_F}\cdot$$
$$\cdot\mathrm{Sp}\left[\begin{array}{l} 2(\boldsymbol{e}_0\boldsymbol{e})^2 J + \\ +c(k_0-k)\left[\begin{array}{l} 2(\boldsymbol{e}_0\boldsymbol{e})(\boldsymbol{\alpha e})(\boldsymbol{\alpha e}_0)+ \\ +2(\boldsymbol{e}_0\boldsymbol{e})\dfrac{(\boldsymbol{\alpha k}_0)(\boldsymbol{\alpha e}_0)(\boldsymbol{\alpha e})(\boldsymbol{\alpha k})}{kk_0}- \\ -(\boldsymbol{\alpha e}_0)(\boldsymbol{\alpha e})2(\boldsymbol{e}_0\boldsymbol{e})- \\ -\dfrac{(\boldsymbol{\alpha k}_0)(\boldsymbol{\alpha e}_0)(\boldsymbol{\alpha e})(\boldsymbol{\alpha k})}{kk_0}2(\boldsymbol{e}_0\boldsymbol{e}) \end{array}\right]\end{array}\right]d\Omega , \quad (6.3.27)$$

The expression in internal square brackets is equally zero. We shall note that the given result excludes the vector form of impulses of the photons, falling and reradiate from the further consideration. It is possible to tell that the zero value of the internal square brackets in (6.3.27) specifies an opportunity and validity of Malus' law from the point of view of quantum electrodynamics since there is no change of the vectors' direction of movement of the photon at its interaction to the electron.

Hence, despite the ambiguity of the physical preconditions discussed earlier in paragraph 6.3.2, the mathematical base of quantum electrodynamics, this at least, at a considered level, adequately describes physical laws; in particular, Malus' law. All further deduction is defined by the absence of the angular scattering of photons, i.e. the reradiated photon always goes in the direction of a falling photon. Thus:

$$d\sigma = e^4 \frac{1}{16E_0{}^2 E} \frac{k}{k_0} \frac{c\partial k}{\partial E_F} \mathrm{Sp}\left[2(e_0 e)^2 J\right] d\Omega = \\ = e^4 \frac{1}{16E_0{}^2 E} \frac{k}{k_0} \frac{c\partial k}{\partial E_F} 8(e_0 e)^2 J d\Omega \tag{6.3.28}$$

Let's find a derivative in the absence of angular scattering of a photon. The energy of the photon-electron system in a final state equals:

$$E_F = ck + E = ck + \sqrt{p^2c^2 + m_0^2c^4} = ck + \sqrt{(k_0 - k)^2 c^2 + E_0^2}$$

The derivative of this expression is:

$$\frac{\partial E_F}{\partial k} = c + \frac{\partial E}{\partial k} = c - \frac{c^2}{E}(k_0 - k) = c\left[\frac{E - cp}{E}\right] = c\frac{E_0}{E}.$$

The law of energy conservation is used as $E - E_0 = c(k_0 - k) = cp$. Substituting in (6.3.28), we shall find:

$$d\sigma = e^4 \frac{J}{2E_0{}^3} \frac{k}{k_0} (e_0 e)^2 d\Omega. \tag{6.3.29}$$

Using and where Θ is an angle of rotation of the polarization plane of quantum after interaction with the electron, we shall obtain:

$$d\sigma = \frac{e^4}{E_0^2}\left[1 + \frac{c(k_0 - k)}{2E_0}\right] \frac{k}{k_0} \cos^2 \Theta d\Omega. \tag{6.3.30}$$

For the further transformation of the formula (6.3.30) it is convenient to enter the classical radius of the electron in system CGS $r_0 = \frac{e^2}{m_0c^2} = \frac{e^2}{E_0} = 2{,}818 \cdot 10^{-13}\ cm$, see paragraph 3.1. As a result, we shall have:

$$d\sigma = r_0^2\left[1+\frac{c(k_0-k)}{2E_0}\right]\frac{k}{k_0}\cos^2\Theta d\Omega. \tag{6.3.31}$$

Furthermore, using the formula (3.1.6) $I = I_0\frac{d\sigma}{r_0^2 d\Omega}$, we obtain a quantum variant of Malus' law:

$$I = I_0\left[1+\frac{c(k_0-k)}{2E_0}\right]\frac{k}{k_0}\cos^2\Theta = I_0\left[1+\frac{\hbar(\omega_0-\omega)}{2m_0c^2}\right]\frac{\omega}{\omega_0}\cos^2\Theta, \tag{6.3.32}$$

where is the intensity of radiations reradiated on electron, and - intensity of radiations falling on electron.

If at the interaction of quantum with the electron its frequency does not vary $\omega_0 = \omega$, we obtain Malus' law (6.2.3) in the classical form.

The formula (6.3.32) is obtained as a result of strict mathematical transformations; the physical problems of a deduction (see paragraph 6.3.2) have been discussed earlier. In the basis of a deduction (6.3.32) there lays the method of small perturbations [4]; therefore, the formula (6.3.32) is essentially inexact. Its form specifies the expansion into a series of some function dependent on the frequency of photons.

A change of reradiated photon frequency appreciably influences the intensity of light which has left the analyzer if the second addend in square brackets (6.3.32) makes approximately 10% from the unit. Under the condition of a difference of wave lengths of photon falling on the electron and being reradiated, it is possible to find from a ratio $\frac{1}{\lambda_0}-\frac{1}{\lambda}\approx\frac{0.2}{\Lambda}$, where is Compton's length of a wave, h - Plank's constant.

Thus, an appreciable influence of change of a photon frequency at the passage of light through the analyzer begins only in the region of rigid X-ray radiation.

Let's transform (6.3.32) to a form:

$$\mu = \frac{I}{I_0} = \left[1+\frac{1}{2}\xi(1-\gamma)\right]\gamma\cos^2\Theta.$$

where and value $\gamma = \frac{\omega}{\omega_0}$. Naturally, $\gamma \le 1$, since part of the photon energy will be transformed to energy of the electron movement.

Let's assume that the electron interact with a photon with Compton's lengths of a wave. Hence, and $\mu = \left(\frac{3-\gamma}{2}\right)\gamma\cos^2\Theta$.

In figure 6.3.4 the calculation of dependence is shown at and $\mu = 0.8$.

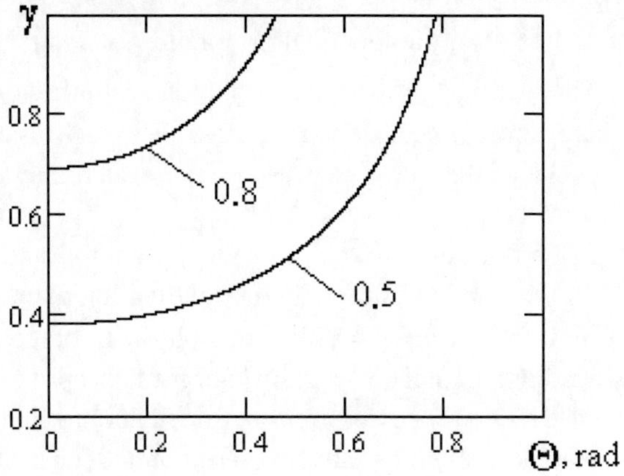

Figure 6.3.4.

From the figure it is visible that a photon directly reradiated on the electron, in the case of constancy μ, is a proportional dependence between the energy reradiated from a photon and an angle of rotation of its polarization plane.

6.3.4. Formula of Klein – Nishina

The formula (6.3.32) found in the previous paragraph describes the directly reradiated photons on the electron without change of direction. However, the light scattering, i.e. a deviation of photon from an initial direction after interaction with electron, is also of interest.

The formula of Klein - Nishina [18] allows us to calculate the differential effective section of quantums scattering on the electron dependent on the angle of scattering θ.

$$d\sigma = \frac{1}{2} r_0^2 \frac{\omega^2}{\omega_0^2} \left(\frac{\omega_0}{\omega} + \frac{\omega}{\omega_0} - \sin^2 \theta \right) d\Omega . \tag{6.3.33}$$

Experiments will confirm the formula (6.3.33) [23].

It is also possible to find the similar formula expressing the differential effective section of the photon scattering on the electron through an angle of rotation of photon polarization plane Θ at its interaction. This formula is also frequently named Klein - Nishina [46]. For comparison with a deduction of the formula of Malus' law (6.3.32) we shall consider a deduction of the Klein - Nishina formula in more detail.

Up to the formula (6.3.27) the deduction of the Klein - Nishina formula completely repeats the deduction of Malus' law.

We shall consider in more detail the formula (6.3.27).

The expression in internal square brackets (6.3.27) has a result equal to zero in connection with the following circumstance. The found differential effective section (6.3.27) generally determines the probability of the rotation of the photon polarization plane on some angle at scattering of the photon on electron or at its direct reradiation. The internal square bracket characterizes the similarity of the probability of rotation of photon polarization plane, as an angle and on angle $-\Theta$; also, the identical probability of scattering of the photon on angle and on angle $-\theta$. The identical probability of photon scattering on angle and on angle results in the internal square bracket being equal to zero. We shall assume that the photon scattering on the electron in one of the sides – namely, on a negative angle - is impossible; i.e. this probability is equal to zero.

For the reception of differential effective section of the photon and electron interaction which take into account both the rotation of the photon polarization plane and its scattering on some positive angle θ , it is necessary to use some physical preconditions (condition). Scattering in one side is accompanied also by the rotation of the photon polarization plane only in one direction since these two processes are functionally connected among themselves.

In internal square brackets (6.3.27) there are two types of addends. The first and the third addends depend only on the angle of rotation of the photon polarization plane. The second and the fourth addends depend on the angle of rotation of the photon polarization plane, and on the angle of photon scattering, and hence characterize the change of its impulse.

We used the strict mathematical formula for transformations to the first and the second addends. Both addends in the used formula are equal and characterize only the rotation of photon polarization plane on angle and $-\Theta$. Equal addends are defined by the property $\cos\Theta = \cos(-\Theta)$. Therefore, for the account of rotation of the photon polarization plane only in one direction (and, hence, one-sided scattering) we used a physical (but not strictly mathematical) condition $2(\boldsymbol{e}_0\boldsymbol{e}) = (\boldsymbol{\alpha e}_0)(\boldsymbol{\alpha e})$. The factor 2 allows us to keep the general normalization of probabilities of photon scattering; ignoring in the formula (6.3.27) the scattering (and, hence, the rotation of photon polarization plane) in one of the sides.

As a result of using $2(\boldsymbol{e}_0\boldsymbol{e}) = (\boldsymbol{\alpha e}_0)(\boldsymbol{\alpha e})$, we shall find:

$$d\sigma = e^4 \frac{1}{16E_0{}^2 E} \frac{k}{k_0} \frac{c\partial k}{\partial E_F} \cdot$$

$$\cdot \mathrm{Sp}\left[\begin{array}{l} 2(\boldsymbol{e}_0\boldsymbol{e})^2 J + \\ +c(k_0 - k)\left[\begin{array}{l} 1+ \\ +\dfrac{(\boldsymbol{\alpha k}_0)(\boldsymbol{\alpha e}_0)(\boldsymbol{\alpha e})(\boldsymbol{\alpha k})(\boldsymbol{\alpha e}_0)(\boldsymbol{\alpha e})}{kk_0} - \\ -(\boldsymbol{\alpha e}_0)(\boldsymbol{\alpha e})2(\boldsymbol{e}_0\boldsymbol{e}) - \\ -\dfrac{(\boldsymbol{\alpha k}_0)(\boldsymbol{\alpha e}_0)(\boldsymbol{\alpha e})(\boldsymbol{\alpha k})}{kk_0} 2(\boldsymbol{e}_0\boldsymbol{e}) \end{array}\right] \end{array}\right] d\Omega$$

The given formula has been found in [46]. Furthermore, with the summation of the second and the fourth addends, and using the full formula (further transformations have a strictly mathematical character), we shall obtain:

$$d\sigma = e^4 \frac{1}{16E_0{}^2 E} \frac{k}{k_0} \frac{c\partial k}{\partial E_F} \cdot$$

$$\cdot \mathrm{Sp}\left[\begin{matrix} 2(\boldsymbol{e}_0\boldsymbol{e})^2 J + \\ + c(k_0 - k)\left[\begin{matrix} 1 - (\boldsymbol{\alpha e}_0)(\boldsymbol{\alpha e})2(\boldsymbol{e}_0\boldsymbol{e}) - \\ - \dfrac{(\boldsymbol{\alpha k}_0)(\boldsymbol{\alpha e}_0)(\boldsymbol{\alpha e})(\boldsymbol{\alpha k})(\boldsymbol{\alpha e})(\boldsymbol{\alpha e}_0)}{kk_0} \end{matrix}\right] \end{matrix}\right] d\Omega \quad (6.3.34)$$

The formula (6.3.34) can be received, proceeding from (6.3.27), more easily. The second addend in the internal square brackets (6.3.27), antisymmetrical to the fourth, is liquidated, i.e. we assume the unidirectional rotation of the polarization plane and photon scattering. In the first and the fourth addends a physical condition is used, that again, with the account $(\boldsymbol{\alpha e})(\boldsymbol{\alpha e}_0) = (\boldsymbol{\alpha e}_0)(\boldsymbol{\alpha e})$, results in (6.3.34).

Once again, we shall emphasize one of the key moments of the deduction of the Klein - Nishina formula. The physical condition allows us to keep the general normalization of probabilities of a photon scattering while ignoring in the formula (6.3.27) its scattering in one of the sides. Using $J = 2E_0 + c(k_0 - k)$, from (6.3.34) we shall find:

$$d\sigma = e^4 \frac{1}{16E_0{}^2 E} \frac{k}{k_0} \frac{c\partial k}{\partial E_F} \cdot$$

$$\cdot \mathrm{Sp}\left[\begin{matrix} 4(\boldsymbol{e}_0\boldsymbol{e})^2 E_0 + 2(\boldsymbol{e}_0\boldsymbol{e})^2 c(k_0 - k) + c(k_0 - k) - \\ - 2(\boldsymbol{e}_0\boldsymbol{e}) c(k_0 - k)(\boldsymbol{\alpha e}_0)(\boldsymbol{\alpha e}) - \\ - c\dfrac{k_0 - k}{kk_0}[(\boldsymbol{\alpha k}_0)(\boldsymbol{\alpha e}_0)(\boldsymbol{\alpha e})(\boldsymbol{\alpha k})(\boldsymbol{\alpha e})(\boldsymbol{\alpha e}_0)] \end{matrix}\right] d\Omega .$$

Further taking into account $\frac{1}{4}\mathrm{Sp}[(\boldsymbol{\alpha e}_0)(\boldsymbol{\alpha e})] = (\boldsymbol{e}_0\boldsymbol{e})$, we have:

$$d\sigma = e^4 \frac{1}{16 E_0{}^2 E} \frac{k}{k_0} \frac{c\partial k}{\partial E_F} \cdot$$

$$\cdot \begin{bmatrix} 16(\boldsymbol{e}_0\boldsymbol{e})^2 E_0 + 4c(k_0 - k) - \\ - c\dfrac{k_0 - k}{k k_0} \mathrm{Sp}[(\boldsymbol{\alpha k}_0)(\boldsymbol{\alpha e}_0)(\boldsymbol{\alpha e})(\boldsymbol{\alpha k})(\boldsymbol{\alpha e})(\boldsymbol{\alpha e}_0)] \end{bmatrix} d\Omega .$$

Taking into account (using cyclic transposition and $(\boldsymbol{\alpha e}_0)^2 = (\boldsymbol{\alpha e})^2 = 1$), that

$$-\frac{1}{4}\mathrm{Sp}[(\boldsymbol{\alpha k}_0)(\boldsymbol{\alpha e}_0)(\boldsymbol{\alpha e})(\boldsymbol{\alpha k})(\boldsymbol{\alpha e})(\boldsymbol{\alpha e}_0)] =$$
$$= -\frac{1}{4}\mathrm{Sp}[(\boldsymbol{\alpha e}_0)(\boldsymbol{\alpha k}_0)(\boldsymbol{\alpha e})(\boldsymbol{\alpha e})(\boldsymbol{\alpha k})(\boldsymbol{\alpha e}_0)] = -\frac{1}{4}\mathrm{Sp}[(\boldsymbol{\alpha k}_0)(\boldsymbol{\alpha k})] = -(\boldsymbol{k}_0\boldsymbol{k}),$$

we shall obtain:

$$d\sigma = e^4 \frac{1}{16 E_0{}^2 E} \frac{k}{k_0} \frac{c\partial k}{\partial E_F} \begin{bmatrix} 16(\boldsymbol{e}_0\boldsymbol{e})^2 E_0 + \\ + 4c(k_0 - k) + c\dfrac{k_0 - k}{k k_0}[-4(\boldsymbol{k}_0\boldsymbol{k})] \end{bmatrix} d\Omega = \tag{6.3.35}$$
$$= e^4 \frac{1}{4 E_0{}^2 E} \frac{k}{k_0} \frac{c\partial k}{\partial E_F} \left[4E_0(\boldsymbol{e}_0\boldsymbol{e})^2 + c\frac{k_0 - k}{k k_0}(k_0 k - (\boldsymbol{k}_0\boldsymbol{k})) \right] d\Omega$$

Furthermore, we use the formula (4.2.28) of a change of a photon frequency at Compton's effect as:

$$k = \frac{k_0}{1 + \dfrac{k_0}{m_0 c}(1 - \cos\theta)}. \tag{6.3.36}$$

where θ is the angle of deviation of a scattered photon from an initial direction. In (6.3.36) over-designations: and $p_0^{/} = k$ are made necessary.

After simple transformations, we shall find:

$$k_0 k - (\boldsymbol{k}_0\boldsymbol{k}) = k_0 k(1 - \cos\theta) = m_0 c(k_0 - k). \tag{6.3.37}$$

Hence, the formula (6.3.35) is of the form:

$$d\sigma = e^4 \frac{1}{4{E_0}^2 E}\frac{k}{k_0}\frac{c\partial k}{\partial E_F}\left[4E_0(\boldsymbol{e}_0\boldsymbol{e})^2 + m_0c^2\frac{(k_0-k)^2}{kk_0}\right]d\Omega =$$
$$= e^4 \frac{1}{4E_0 E}\frac{k}{k_0}\frac{c\partial k}{\partial E_F}\left[4(\boldsymbol{e}_0\boldsymbol{e})^2 + \frac{(k_0-k)^2}{kk_0}\right]d\Omega = \tag{6.3.38}$$
$$= e^4 \frac{1}{4E_0 E}\frac{k}{k_0}\frac{c\partial k}{\partial E_F}\left(\frac{k_0}{k} + \frac{k}{k_0} - 2 + 4\cos^2\Theta\right)d\Omega$$

Let's find a derivative at the opportunity of the angular scattering of a photon. The energy of the photon-electron system in the final state equals:

$$E_F = ck + E = ck + \sqrt{p^2c^2 + m_0^2c^4} = ck + \sqrt{(\boldsymbol{k}_0 - \boldsymbol{k})^2c^2 + E_0^2} =$$
$$= ck + \sqrt{k_0^2c^2 - 2k_0kc^2\cos\theta + k^2c^2 + E_0^2}$$

A derivative of this expression:

$$\frac{\partial E_F}{\partial k} = c + \frac{\partial E}{\partial k} = c - \frac{c^2}{E}(k_0\cos\theta - k) = c\left[\frac{Ek - c(k_0k\cos\theta - k^2)}{Ek}\right] =$$
$$= c\left[\frac{Ek - c(-m_0c(k_0-k) + k_0k - k^2)}{Ek}\right] = ck_0\left[\frac{-k_0c + E + kc}{Ek}\right] = \frac{ck_0E_0}{Ek}$$

At the deduction the formula (6.3.37) and the law of energy conservation is used. Substituting in (6.3.38) $\dfrac{\partial k}{\partial E_F} = \dfrac{Ek}{ck_0E_0}$, we shall find:

$$d\sigma = \frac{1}{4}r_0^2\frac{k^2}{k_0^2}\left(\frac{k_0}{k} + \frac{k}{k_0} - 2 + 4\cos^2\Theta\right)d\Omega =$$
$$= \frac{1}{4}r_0^2\frac{\omega^2}{\omega_0^2}\left(\frac{\omega_0}{\omega} + \frac{\omega}{\omega_0} - 2 + 4\cos^2\Theta\right)d\Omega \tag{6.3.39}$$

Let's note that the formula (6.3.39) is deduced with use of the formula (6.3.36), i.e. at the existence of angular scattering of a photon. Therefore, it basically characterizes Compton's effect.

As was already specified, between an angle of scattering θ and the angle of rotation of the photon polarization plane Θ in Compton's effect there is a functional constraint. It can be found having equated the differential effective section of the process expressed through the angle of scattering (6.3.33) and the section of the process expressed through the angle of rotation of the photon polarization plane (6.3.39). As a result, we shall find $\left(\frac{1}{\gamma}+\gamma-1\right)+\cos 2\theta = 2\cos 2\Theta$, where $\gamma = \frac{\omega}{\omega_0}$. In figure 6.3.5 dependence is shown at - there is no change of frequency of a photon, and $\gamma = 0.5$.

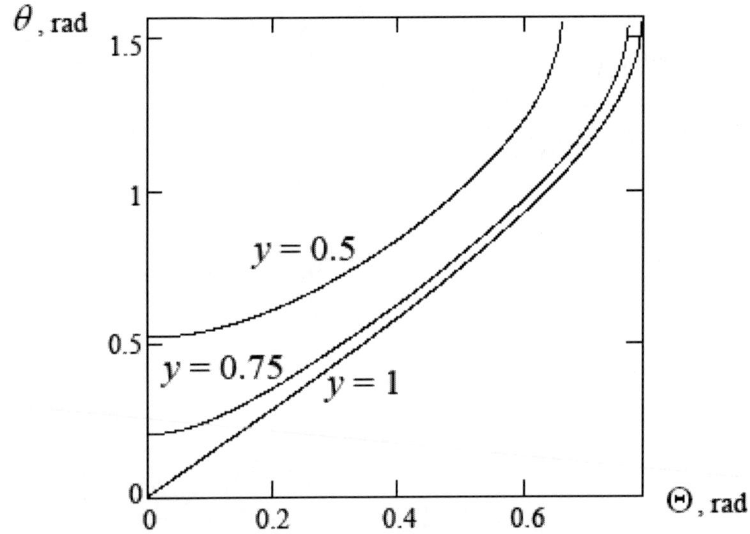

Figure 6.3.5.

From the figure it is visible that at the increase in the angle of scattering θ, the angle of rotation of the photon polarization plane Θ also increases.

6.3.5. Physical Reason of Variation of the Photon Frequency After Its Direct Reradiation on the Electron

The formula of Malus' law (6.3.32) shows that at the rotation of the photon polarization plane after it is reradiated on the electron, the energy of the photon can decrease. In particular, it can be transferred to the atom, i.e. transform in heat. However, the physical mechanism of the change of the photon frequency is not clear.

For an explanation of the physical reason for the variation of the photon frequency after its direct reradiation on the electron without change of direction, we shall consider in more detail the formula (6.3.27) for the differential effective section of interaction of the photon and electron. The second addend in external square brackets (6.3.27) appeared equal to zero. Keeping this addend formally shall lead us to some transformations (6.3.27).

Using rules of work with spurs [4]: and

$$\frac{1}{4}\mathrm{Sp}[(\boldsymbol{\alpha k}_0)(\boldsymbol{\alpha e}_0)(\boldsymbol{\alpha e})(\boldsymbol{\alpha k})]=$$
$$=[(\boldsymbol{k}_0\boldsymbol{e}_0)(\boldsymbol{ek})+(\boldsymbol{k}_0\boldsymbol{k})(\boldsymbol{e}_0\boldsymbol{e})-(\boldsymbol{k}_0\boldsymbol{e})(\boldsymbol{e}_0\boldsymbol{k})]=[(\boldsymbol{k}_0\boldsymbol{k})(\boldsymbol{e}_0\boldsymbol{e})-(\boldsymbol{k}_0\boldsymbol{e})(\boldsymbol{e}_0\boldsymbol{k})]=,$$
$$=[\boldsymbol{k}_0\times\boldsymbol{e}_0][\boldsymbol{k}\times\boldsymbol{e}]=k_0k(\boldsymbol{l}_0\boldsymbol{l})=k_0k(\boldsymbol{e}_0\boldsymbol{e})$$

and also the formula of vector algebra [21] where and there is unit vectors, we shall find:

$$d\sigma=e^4\frac{1}{16E_0^{\ 3}}\frac{k}{k_0}\left[\begin{matrix}8(\boldsymbol{e}_0\boldsymbol{e})^2J+\\ +c(k_0-k)\left[\begin{matrix}8(\boldsymbol{e}_0\boldsymbol{e})^2+8(\boldsymbol{e}_0\boldsymbol{e})^2-\\ -8(\boldsymbol{e}_0\boldsymbol{e})^2-8(\boldsymbol{e}_0\boldsymbol{e})^2\end{matrix}\right]\end{matrix}\right]d\Omega= \tag{6.3.40}$$
$$=e^4\frac{1}{2E_0^{\ 3}}(\boldsymbol{e}_0\boldsymbol{e})^2\frac{k}{k_0}[J+c(k_0-k)[2-2]]d\Omega$$

It is used, also, that at the direct reradiation of photons the derivative in Malus' law is $\frac{\partial k}{\partial E_F}=\frac{E}{cE_0}$. Besides, it is taken into account and is the right three vectors. Rotation of the photon polarization plane is carried out around vectors of the impulse and $\boldsymbol{k}$; therefore, $(\boldsymbol{l}_0\boldsymbol{l})=(\boldsymbol{e}_0\boldsymbol{e})$.

Substituting in (6.3.40) value $J=2E_0+c(k_0-k)$, we shall find:

$$d\sigma=e^4\frac{1}{E_0^{\ 3}}(\boldsymbol{e}_0\boldsymbol{e})^2\frac{k}{k_0}\left[E_0+\frac{c(k_0-k)}{2}+c(k_0-k)[1-1]\right]d\Omega. \tag{6.3.41}$$

Transitioning from the impulses of photons to their frequencies, we obtain:

$$d\sigma=e^4\frac{1}{E_0^{\ 3}}(\boldsymbol{e}_0\boldsymbol{e})^2\frac{\omega}{\omega_0}\left[E_0+\hbar\Delta\omega[1-1]+\frac{\hbar(\omega_0-\omega)}{2}\right]d\Omega, \tag{6.3.42}$$

where $\Delta\omega=\omega_0-\omega$.

The formula (6.3.42) allows us to draw the conclusion that the electron is not in a motionless state before interaction with a photon. Its energy varies in limits $E_0\pm\hbar\Delta\omega$. This fluctuation in free space can occur only due to the influence of the vacuum energy on the electron. For example, it is similar to the influence of vacuum energy on orbital electrons in atom; there is a thin structure of Lamb's spectrum [4]. Rather, conditional fluctuations of the electron energy are shown in figure 6.3.6.

After the absorption of the photon with energy, and the transition of the electron to the intermediate state, the fluctuating electron then, according to the diagram in figure 6.3.6, transitions to a final state with energy *E*. Thus, the photon is radiated by energy. Energy *E* is defined by the mean square position on space fluctuating electron. The shift of the electron energy $E - E_0$, like Lamb's spectral shifts, is arising due to the energy of vacuum. The second variant, in conformity with figure 6.3.6, consists of the electron at first radiating a photon and transitioning to an intermediate state E_{II}. After that, the electron absorbs a photon energy and transitions to final state *E*. Levels of intermediate states are shown conditionally for presentation in figure 6.3.6. Apparently, these states cannot be placed on the linear energy scale.

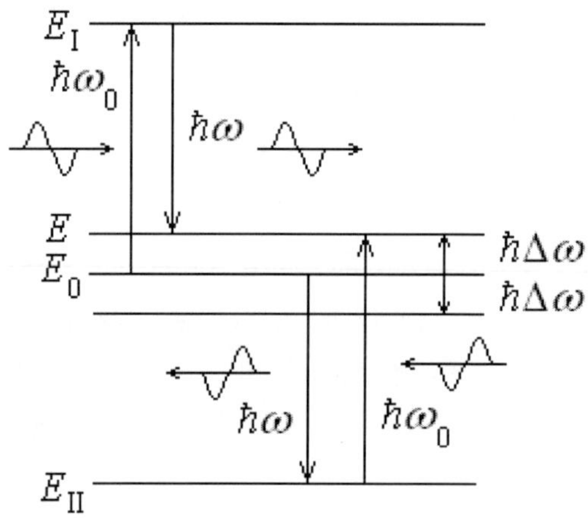

Figure 6.3.6.

The transition of the electron after interaction with a photon on the energy level *E* is defined by the correctness of the energy conservation law before and after the interaction of a photon with the electron. Whether the transition of the electron is to the level $E_0 - \Delta\omega$, to this question only an experiment can possibly provide an answer. However, the existence of the combinational scattering [51] at the interaction of an electromagnetic wave on oscillating charged structures of atoms and molecules allows us to assume that the level can be important at the interaction of the photon with the electron.

Thus, it is shown that, despite some of the ambiguity of physical preconditions existing now, the mathematical base of quantum electrodynamics, at least, at a considered level, adequately describes the physical laws; in particular, Malus' law.

To use Malus' law (6.2.3) based on the principles of classical electrodynamics does not completely take into account all the effects which can occur at the passage of light through a polarizer-analyzer system. The phenomenon of the possible change of light frequency in particular drops out, especially at high-frequency; for example, in the region of X-ray radiation.

Malus' law is based on the principles of quantum electrodynamics (6.3.32), allowing us to estimate the influence of the change of light frequency on the intensity of the passage of light through a polarizer-analyzer system. This effect needs taking into account, for example, in the development of X-ray lasers.

The variation of a reradiated photon frequency, without a change of direction, on the electron occurs due to the interaction of the electron with the energy of vacuum.

The formula for the differential effective section of the scattering of a photon on the electron (6.3.39) has a narrower range of use (connected only with Compton's effect), than the formula (6.3.30) resulting in Malus' law for X-ray radiation (6.3.32).

6.4. Quantum Theory of Optical Activity

The classical approach to the research of optical activity in the paragraph 2.4 has been applied. The additional information on this phenomenon can be received at the quantum approach. The satisfactory quantum theory of optical activity of environments for the first time has been developed by a pupil of N. Bohr; the Belgian physicist L. Rosenfeld in 1928 [62]. This theory, which does not have a relativistic character, is based on the standard approach of research of the influence of electromagnetic field on quantum system - a molecular optically active substance with the help of the theory of small perturbations. The major feature of the molecule of an optically active substance is its asymmetry; i.e. this molecule should not have a plane or center of symmetry [52].

The relativistic variant of the description of optical activity allows us to receive additional information on the process.

6.4.1. Non-relativistic Variant

Let's consider the interaction of quantums of electromagnetic radiation and the molecules of the optically active substance. For the process of interaction of electromagnetic radiation and the molecule of the optically active substance, Hamilton's operator, it is possible to present as the sum:

$$\hat{H} = \hat{H}_0 + \hat{V}(t), \tag{6.4.1}$$

where is the operator determining the state of the molecule without influence on it, i.e. not dependent on time addend, - the time-dependent operator of interaction of the molecule and photon.

Schrodinger's equation for the considered process can be presented as:

$$i\hbar\frac{\partial\Psi}{\partial t} = \left(\hat{H}_0 + \hat{V}\right)\Psi, \tag{6.4.2}$$

where ψ is the wave function of process of the light quantum interaction and the molecule of optically active substance, - Planck's reduced constant. Let's search the solution for the equation (6.4.2), having presented the wave function of the process of interaction as:

$$\Psi = \sum_k a_k(t)\Psi_k^{(0)}, \tag{6.4.3}$$

where is the wave function of a stationary state at number *k*, - factors of expansion into series of perturbing wave function Ψ on stationary states with wave function $\Psi_k^{(0)}$. Above in brackets is the order of approximation of the solution of the equation (6.4.2).
Having substituted (6.4.3) in (6.4.2), we shall find:

$$\sum_k \left(i\hbar\Psi_k^{(0)} \frac{\partial a_k(t)}{\partial t} + i\hbar a_k(t)\frac{\partial \Psi_k^{(0)}}{\partial t} \right) = \hat{H}_0 \sum_k a_k(t)\Psi_k^{(0)} + \hat{V}\sum_k a_k(t)\Psi_k^{(0)}, \tag{6.4.4}$$

Taking into account that the stationary wave function of the zero order (i.e. not perturbing) satisfies the equation:

$$i\hbar\frac{\partial \Psi_k^{(0)}}{\partial t} = \hat{H}_0\Psi_k^{(0)}, \tag{6.4.5}$$

we shall find:

$$i\hbar\sum_k \Psi_k^{(0)}\frac{\partial a_k(t)}{\partial t} = \sum_k a_k(t)\hat{V}\Psi_k^{(0)}, \tag{6.4.6}$$

Let's multiply the equation (6.4.6) at the left on the conjugate wave function and we shall integrate the set of coordinates of quantum system *q*. As a result, we shall obtain:

$$i\hbar\sum_k \frac{\partial a_k(t)}{\partial t}\int \Psi_m^{(0)*}\Psi_k^{(0)}dq = \hbar\sum_k \frac{\partial a_k(t)}{\partial t}\delta_{mk} = \sum_k a_k(t)\int \Psi_m^{(0)*}\hat{V}\Psi_k^{(0)}dq,$$

$$i\hbar\frac{\partial a_m(t)}{\partial t} = \sum_k a_k(t)V_{mk}(t), \tag{6.4.7}$$

where at and at $m \neq k$, the values there are matrix elements of perturbation from a non-perturbing state *k* up to the perturbing state *m* into which the exponential time factor is allocated.

Let's lead the over-designation of indexes. Let the index *n* there correspond to non-perturbing state, an index *k* - the perturbing state. Thus, and at $k \neq n$. Being limited to the first approximation of the solution (6.4.3), we assume that the factor $a_{kn} = a_{kn}^{(0)} + a_{kn}^{(1)} = a_{kn}^{(1)}$. The second bottom index specifies how the correction (i.e. to a state *n*) is calculated from non-perturbing states.

Hence:

$$i\hbar\frac{\partial a_{kn}^{(1)}(t)}{\partial t}=a_n^{(0)}V_{kn}(t). \quad (6.4.8)$$

For the convenience of a designation of quantum transitions we shall write down (6.4.3) as:

$$\Psi_n=\sum_k a_{kn}(t)\Psi_k^{(0)}, \quad (6.4.9)$$

and (6.4.8), taking into account $a_n^{(0)}=1$, as:

$$i\hbar\frac{\partial a_{kn}^{(1)}(t)}{\partial t}=V_{kn}(t)=V_{kn}e^{\frac{i}{\hbar}W_{kn}t}. \quad (6.4.10)$$

where is a difference energy of molecules in k and n states. The time factor of a matrix element of perturbation in (6.4.10) is allocated separately.

Up until now we have not paid attention to the physical nature of perturbation. Believing that perturbation is carried out due to light quantum, we shall consider periodic perturbation as:

$$\hat{V}=\hat{V}_+e^{i\omega t}+\hat{V}_-e^{-i\omega t}, \quad (6.4.11)$$

where the first addend characterizes the influence on the molecule of quantum frequency ω right-circular polarization, and the second addend – of quantum frequency ω left-circular polarization. At perturbation looks like $\hat{V}=\hat{V}_+\left(e^{i\omega t}+e^{-i\omega t}\right)=2\hat{V}_+\cos\omega t$, i.e. in this case the influence on a molecule is carried out by quantum with linear polarization which can be presented as the sum of two quantums with right and left-circular polarizations.

Using (6.4.11), we shall find matrix elements of perturbation:

$$V_{kn}(t)=\int\Psi_k^{(0)*}\hat{V}_+e^{i\omega t}\Psi_n^{(0)}dq+\int\Psi_k^{(0)*}\hat{V}_-e^{i\omega t}\Psi_n^{(0)}dq=$$
$$=V_{+kn}e^{\frac{i}{\hbar}W_{kn}t+i\omega t}+V_{-kn}e^{\frac{i}{\hbar}W_{kn}t-i\omega t}, \quad (6.4.12)$$

where and are the matrix elements of perturbations due to quantums' various circular polarizations.

Substituting (6.4.12) in (6.4.10) and integrating on time (6.4.10) in limits from 0 up to t, we find:

$$a_{kn}^{(1)}(t)=-\frac{i}{\hbar}\int V_{kn}(t)dt=-\left(\frac{V_{+kn}e^{\frac{i}{\hbar}(W_{kn}+E)t}}{W_{kn}+E}+\frac{V_{-kn}e^{\frac{i}{\hbar}(W_{kn}-E)t}}{W_{kn}-E}\right), \qquad (6.4.13)$$

where is the energy of the quantum falling on a molecule.

At integration, we assume [4] that at the factor of perturbation was absent, i.e. $\widehat{V}\equiv 0$.

Hence, the correction of the first order to the wave function of the molecule, satisfying the equation (6.4.2), looks like:

$$\Psi_n^{(1)}=-\sum_k \Psi_k^{(0)}\left(\frac{V_{+kn}e^{\frac{i}{\hbar}(W_{kn}+E)t}}{W_{kn}+E}+\frac{V_{-kn}e^{\frac{i}{\hbar}(W_{kn}-E)t}}{W_{kn}-E}\right). \qquad (6.4.14)$$

Perturbing influence on a molecule of an electromagnetic field of quantum is connected with the appearance of induced electric and magnetic quantum dipole moments at a molecule.

Let's find matrix elements of these dipole moments. As a first approximation for an electric dipole [53]:

$$\boldsymbol{p}_n(t)=\boldsymbol{p}_n^{(0)}(t)+\boldsymbol{p}_n^{(1)}(t)=\boldsymbol{p}_n^{(1)}(t). \qquad (6.4.15)$$

We believe that initially the molecule has no dipole moment $\boldsymbol{p}_n^{(0)}(t)=0$.

As a first approximation:

$$\boldsymbol{p}_n^{(1)}(t)=\sum_k \mathrm{R}\int_q \Psi_k^{(0)*}\hat{\boldsymbol{p}}\Psi_n^{(1)}dq=$$

$$=-\sum_k \mathrm{R}\int_q \Psi_k^{(0)*}\hat{\boldsymbol{p}}\Psi_k^{(0)}dq\left(\frac{V_{+kn}e^{\frac{i}{\hbar}(W_{kn}+E)t}}{W_{kn}+E}+\frac{V_{-kn}e^{\frac{i}{\hbar}(W_{kn}-E)t}}{W_{kn}-E}\right)=\cdot \qquad (6.4.16)$$

$$=-\sum_k \mathrm{R}\boldsymbol{p}_{kn}\left(\frac{V_{+kn}e^{\frac{i}{\hbar}(W_{kn}+E)t}}{W_{kn}+E}+\frac{V_{-kn}e^{\frac{i}{\hbar}(W_{kn}-E)t}}{W_{kn}-E}\right)$$

where $\boldsymbol{p}_{kn} = \int_q \Psi_k^{(0)*} \hat{\boldsymbol{p}} \Psi_k^{(0)} dq$. The index n in this case marks an initial stationary state of the molecule concerning which there arises a dipole moment. Integration goes on volume q molecules, R - the operator of allocation of the real part. The operators' electric and magnetic dipole moments in the Cartesian coordinates look like [52]:

$$\hat{\boldsymbol{p}} = e(\boldsymbol{i}x + \boldsymbol{j}y + \boldsymbol{k}z),$$

$$\hat{\boldsymbol{m}} = -i\frac{e\hbar}{2m_e c}\left\{\boldsymbol{i}\left(y\frac{\partial}{\partial z} - z\frac{\partial}{\partial y}\right) + \boldsymbol{j}\left(z\frac{\partial}{\partial x} - x\frac{\partial}{\partial z}\right) + \boldsymbol{k}\left(x\frac{\partial}{\partial y} - y\frac{\partial}{\partial x}\right)\right\},$$

where standard designations are used: e is a charge of the electron, - its mass, c - speed of light in vacuum, - unit vectors.

Transforming (6.4.16) with the account (6.4.15), we shall find:

$$\boldsymbol{p}_n(t) = -\sum_k \mathrm{R}\frac{\boldsymbol{p}_{kn}}{W_{kn}^2 - E^2}\left(V_{+kn}(W_{kn} - E)e^{\frac{i}{\hbar}(W_{kn}+E)t} + V_{-kn}(W_{kn} + E)e^{\frac{i}{\hbar}(W_{kn}-E)t}\right) \quad (6.4.17)$$

Similarly for a magnetic dipole:

$$\boldsymbol{m}_n(t) = -\sum_k \mathrm{R}\boldsymbol{m}_{kn}\left(\frac{V_{+kn}e^{\frac{i}{\hbar}(W_{kn}+E)t}}{W_{kn} + E} + \frac{V_{-kn}e^{\frac{i}{\hbar}(W_{kn}-E)t}}{W_{kn} - E}\right) = \quad (6.4.18)$$

$$= -\sum_k \mathrm{R}\frac{\boldsymbol{m}_{kn}}{W_{kn}^2 - E^2}\left(V_{+kn}(W_{kn} - E)e^{\frac{i}{\hbar}(W_{kn}+E)t} + V_{-kn}(W_{kn} + E)e^{\frac{i}{\hbar}(W_{kn}-E)t}\right)$$

where $\boldsymbol{m}_{kn} = \int_q \Psi_k^{(0)*} \hat{\boldsymbol{m}} \Psi_k^{(0)} dq$.

Let's assume that the influence of electromagnetic radiation on a molecule linearly depends on strength from electric and magnetic fields. It is correct only for radiation of small intensity. Besides photons of the right and left-circular polarization, they act equally on the molecule, for the occurrence of electric and magnetic dipole moments.

Therefore, matrix factors of perturbation, i.e. reaction of the molecule to electromagnetic influence can be presented as:

$$V_{skn}e^{\frac{i}{\hbar}(W_{kn}+sE)t} = -(\boldsymbol{p}_{kn}\boldsymbol{E} + \boldsymbol{m}_{kn}\boldsymbol{H})\frac{(W_{kn} + E)}{(W_{kn} - sE)}. \quad (6.4.19)$$

where are spins of photons of the right and left-circular polarization.

Let's transform (6.4.19):

$$V_{skn}\left(W_{kn}-sE\right)e^{\frac{i}{\hbar}\left(W_{kn}+sE\right)t}=-\left(\boldsymbol{p}_{kn}\boldsymbol{E}+\boldsymbol{m}_{kn}\boldsymbol{H}\right)\left(W_{kn}+E\right)=\cdot$$
$$=-\left(\boldsymbol{p}_{kn}\boldsymbol{E}W_{kn}+\boldsymbol{p}_{kn}\boldsymbol{E}E+\boldsymbol{m}_{kn}\boldsymbol{H}W_{kn}+\boldsymbol{m}_{kn}\boldsymbol{H}E\right) \quad (6.4.20)$$

Using a condition of harmonious character of influence on a molecule and $\boldsymbol{H}=\boldsymbol{H}_0 e^{-\frac{iEt}{\hbar}}$, we have and $i\hbar\frac{\partial \boldsymbol{H}}{\partial t}=E\boldsymbol{H}$.

Hence:

$$V_{skn}\left(W_{kn}-sE\right)e^{\frac{i}{\hbar}\left(W_{kn}+sE\right)t}= .$$
$$=-\left(\boldsymbol{p}_{kn}\boldsymbol{E}W_{kn}+i\hbar\boldsymbol{p}_{kn}\frac{\partial \boldsymbol{E}}{\partial t}+\boldsymbol{m}_{kn}\boldsymbol{H}W_{kn}+i\hbar\boldsymbol{m}_{kn}\frac{\partial \boldsymbol{H}}{\partial t}\right) \quad (6.4.21)$$

Substituting in (6.4.17) and (6.4.18) formula (6.4.21), we shall find:

$$\boldsymbol{p}_n=2\sum_k \mathrm{R}\frac{1}{W_{kn}^2-E^2}\boldsymbol{p}_{kn}\left(\boldsymbol{p}_{kn}\boldsymbol{E}W_{kn}+i\hbar\boldsymbol{p}_{kn}\frac{\partial \boldsymbol{E}}{\partial t}+\boldsymbol{m}_{kn}\boldsymbol{H}W_{kn}+i\hbar\boldsymbol{m}_{kn}\frac{\partial \boldsymbol{H}}{\partial t}\right), \quad (6.4.22)$$

$$\boldsymbol{m}_n=2\sum_k \mathrm{R}\frac{1}{W_{kn}^2-E^2}\boldsymbol{m}_{kn}\left(\boldsymbol{p}_{kn}\boldsymbol{E}W_{kn}+i\hbar\boldsymbol{p}_{kn}\frac{\partial \boldsymbol{E}}{\partial t}+\boldsymbol{m}_{kn}\boldsymbol{H}W_{kn}+i\hbar\boldsymbol{m}_{kn}\frac{\partial \boldsymbol{H}}{\partial t}\right) \quad (6.4.23)$$

The values and are real. Therefore, the second addend in brackets (6.4.22) is only imaginary in size and the real part of this value is equal to zero. In (6.4.23) it is neglected by quadrupole the magnetic sizes proportional owing to its small size. There is factor 2 owing to two spin states of a photon.

The real part of size where X is the function $\boldsymbol{m}_k$, is where means an imaginary part X.

Therefore, we shall write down:

$$\boldsymbol{p}_n=\frac{2}{3}\sum_k\frac{1}{W_{kn}^2-E^2}\left(\left(\boldsymbol{p}_{kn}\boldsymbol{p}_{kn}\right)\boldsymbol{E}W_{kn}+\mathrm{R}\left(\boldsymbol{p}_{kn}\boldsymbol{m}_{kn}\right)\boldsymbol{H}W_{kn}-\hbar\,\mathrm{Im}\left(\boldsymbol{p}_{kn}\boldsymbol{m}_{kn}\right)\frac{\partial \boldsymbol{H}}{\partial t}\right) \quad (6.4.24)$$

$$\boldsymbol{m}_n=\frac{2}{3}\sum_k\frac{1}{W_{kn}^2-E^2}\left(\mathrm{R}\left(\boldsymbol{m}_{kn}\boldsymbol{p}_{kn}\right)\boldsymbol{E}W_{kn}-\hbar\,\mathrm{Im}\left(\boldsymbol{m}_{kn}\boldsymbol{p}_{kn}\right)\frac{\partial \boldsymbol{E}}{\partial t}\right)= ,$$
$$=\frac{2}{3}\sum_k\frac{1}{W_{kn}^2-E^2}\left(\mathrm{R}\left(\boldsymbol{p}_{kn}\boldsymbol{m}_{kn}\right)\boldsymbol{E}W_{kn}+\hbar\,\mathrm{Im}\left(\boldsymbol{p}_{kn}\boldsymbol{m}_{kn}\right)\frac{\partial \boldsymbol{E}}{\partial t}\right) \quad (6.4.25)$$

where it is taken into account and $\mathrm{R}(\boldsymbol{m}_{kn}\boldsymbol{p}_{kn}) = \mathrm{R}(\boldsymbol{p}_{kn}\boldsymbol{m}_{kn})$.
The factor 1/3 results from averaging products of vectors and on all directions.

Thus, we receive a reaction of the molecule to the electromagnetic influence (the index of a stationary quantum state *n* is omitted) as, see (2.4.34):

$$\boldsymbol{p} = \alpha \boldsymbol{E} + \gamma \boldsymbol{H} - \frac{\beta}{c}\frac{\partial \boldsymbol{H}}{\partial t} \text{ and } \boldsymbol{m} = \gamma \boldsymbol{E} + \frac{\beta}{c}\frac{\partial \boldsymbol{E}}{\partial t}, \tag{6.4.26}$$

where it is designated:

$$\alpha = \frac{2}{3}\sum_k \frac{\boldsymbol{p}_{kn}\boldsymbol{p}_{kn}W_{kn}}{W_{kn}^2 - E^2} = \frac{2}{3\hbar}\sum_k \frac{\omega_{kn}\boldsymbol{p}_{kn}\boldsymbol{p}_{kn}}{\omega_{kn}^2 - \omega^2}, \tag{6.4.27}$$

$$\beta = \frac{2}{3}\sum_k \frac{\hbar\,\mathrm{Im}(\boldsymbol{p}_{kn}\boldsymbol{m}_{kn})}{W_{kn}^2 - E^2} = \frac{2c}{3\hbar}\sum_k \frac{\mathrm{Im}(\boldsymbol{p}_{kn}\boldsymbol{m}_{kn})}{\omega_{kn}^2 - \omega^2}, \tag{6.4.28}$$

$$\gamma = \frac{2}{3}\sum_k \frac{\mathrm{R}(\boldsymbol{p}_{kn}\boldsymbol{m}_{kn})W_{kn}}{W_{kn}^2 - E^2} = \frac{2}{3\hbar}\sum_k \frac{\omega_{kn}\mathrm{R}(\boldsymbol{p}_{kn}\boldsymbol{m}_{kn})}{\omega_{kn}^2 - \omega^2}. \tag{6.4.29}$$

In formulas (6.4.27) - (6.4.29) is the frequency of a resonance at the interaction of a molecule with radiation on electronic transition $n \to k$, α - polarizability of a molecule. The value:

$$R_{kn} = \mathrm{Im}(\boldsymbol{p}_{kn}\boldsymbol{m}_{kn}) \tag{6.4.30}$$

is called the rotary force of transition $n \to k$. Parameter:

$$\beta = \frac{2c}{3\hbar}\sum_k \frac{R_{kn}}{\omega_{kn}^2 - \omega^2} \tag{6.4.31}$$

called the rotating ability of the molecule (environment).

Let's note two important properties of rotary force. First, $R_{kn} = -R_{nk}$. Rotary force of transition is equal to the rotary force of transition with the minus sign. Secondly, is correct for all states *n*, and summation is realized on all states *k* [52].

In the area of a band of interaction of quantum with the electrons' transition of the molecule by energy $W_{kn} = \hbar\omega_{kn}$, it is necessary to take into account the attenuation of oscillations of the electronic orbit (or, moreover, the increase in width of the excited energy level owing to correlations of Heisenberg's uncertainty). Therefore, for example, for rotating ability of the environment it is necessary to use the formula [52]:

$$\beta = \frac{2c}{3\hbar}\sum_k \frac{R_{kn}}{\omega_{kn}^2 - \omega^2 + i\omega H_{kn}} = \\ = \frac{2c}{3\hbar}\sum_k \frac{R_{kn}\left(\omega_{kn}^2 - \omega^2\right)}{\left(\omega_{kn}^2 - \omega^2\right)^2 + \omega^2 H_{kn}^2} - \frac{2ic}{3\hbar}\sum_k \frac{R_{kn}\omega H_{kn}}{\left(\omega_{kn}^2 - \omega^2\right)^2 + \omega^2 H_{kn}^2}, \quad (6.4.32)$$

where characterizes the half-width frequency of the band of absorption on transition due to the energy increase in the width of the band of the excited level *k*.

Let's note that the first addend in the formula (6.4.32) characterizes a dispersion of the optical rotation, the second - circular dichroism, see paragraph 2.4.1. Near to a band of interaction of the quantum light and molecule the abnormal dispersion of optical rotation takes place.

6.4.2. Relativistic Variant

We shall begin the quantum-relativistic analysis of optical activity with research into the interaction of photons and electrons in a molecule by method of small perturbations. Dirac's equation, except for a form of Hamilton's operator, does not differ from Schrodinger's equation (6.4.2); therefore, results of paragraph 6.4.1 as a whole are applicable to the relativistic analysis of optical activity.

Interaction of a photon and molecule can occur in two ways; figure 6.4.1.

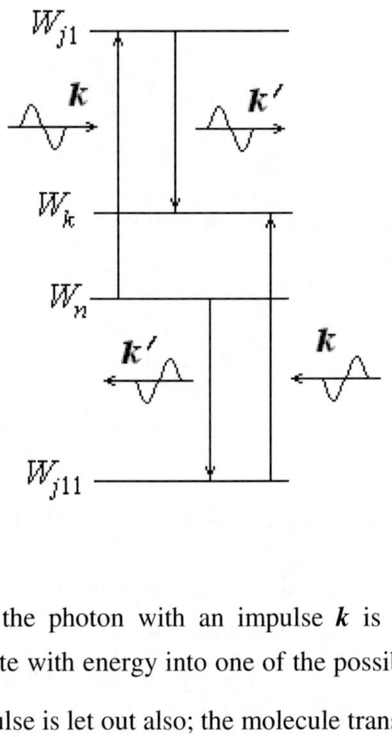

Figure 6.4.1.

The first way is when the photon with an impulse $\boldsymbol{k}$ is absorbed, and the molecule transitions from its initial state with energy into one of the possible intermediate states W_{j1}. Then the photon having impulse is let out also; the molecule transitions into a final state W_k.

The second way is when the molecule lets out a photon and it transitions from initial state with energy into the intermediate state W_{j11}. Then at the transition in a final state the photon **k** is absorbed.

The matrix element of the examined process looks like [18]:

$$V_{kn}(t)=\sum_j\left[\frac{V_{kj}^{/}V_{jn}(t)}{W_n-W_{j1}+\hbar\omega}+\frac{V_{kj}(t)V_{jn}^{/}}{W_n-W_{j11}-\hbar\omega^{/}}\right], \quad (6.4.33)$$

Values are matrix elements of absorption of the photon ***k***, values - matrix elements of radiation of the photon $\boldsymbol{k}^{/}$. The initial state is excluded from summation on *j*. We shall note that the process of perturbation is defined only by the quantum falling on a molecule, i.e. matrix elements without a prime. The radiated quantum can cause only a reaction of radiation [21] which we shall estimate further.

Let's write down for two variants of absorption of quantums by the molecule and $j11\rightarrow k$, figure 6.1.4, the matrix elements of perturbations as:

$$V_{jn}(t)=V_{jn}e^{\frac{i}{\hbar}(W_{j1}-W_n)t} \text{ and } V_{kj}(t)=V_{kj}e^{\frac{i}{\hbar}(W_k-W_{j11})t}. \quad (6.4.34)$$

Inserting (6.4.34) in (6.4.33), we shall find:

$$V_{kn}(t)=\sum_j\left[\frac{V_{kj}^{/}V_{jn}e^{\frac{i}{\hbar}(W_{j1}-W_n)t}}{W_n-W_{j1}+\hbar\omega}+\frac{V_{kj}V_{jn}^{/}e^{\frac{i}{\hbar}(W_k-W_{j11})t}}{W_n-W_{j11}-\hbar\omega^{/}}\right].$$

Assuming that perturbation due to light quantum looks like (6.4.11), we obtain:

$$V_{kn}(t)=\sum_s\sum_j\left[\frac{V_{skj}^{/}V_{sjn}e^{\frac{i}{\hbar}(W_{j1}-W_n+sE)t}}{W_n-W_{j1}+\hbar\omega}+\frac{V_{skj}V_{sjn}^{/}e^{\frac{i}{\hbar}(W_k-W_{j11}+sE)t}}{W_n-W_{j11}-\hbar\omega^{/}}\right], \quad (6.4.35)$$

where $E=\hbar\omega$, are the spins of photons of the right and left-circular polarization.

Factors of the expansion into a series of the wave function as a first approximation of the theory of perturbations look like:

$$a_{kn}^{(1)}(t) = -\frac{i}{\hbar}\int V_{kn}(t)dt = -\sum_s\sum_j \left[\frac{V_{skj}^{/} V_{sjn} e^{\frac{i}{\hbar}\left(W_{j1}-W_n+sE\right)t}}{\left(W_n - W_{j1} + \hbar\omega\right)\left(W_{j1} - W_n + sE\right)} + \atop + \frac{V_{skj} V_{sjn}^{/} e^{\frac{i}{\hbar}\left(W_k - W_{j11}+sE\right)t}}{\left(W_n - W_{j11} - \hbar\omega^{/}\right)\left(W_k - W_{j11} + sE\right)} \right]. \quad (6.4.36)$$

Let's transform the first addend in the formula (6.4.36), having multiplied the numerator and denominator on value $W_n + W_{j1} + \hbar\omega$. Transformations to the denominator look like:

$$\left(W_n - W_{j1} + \hbar\omega\right)\left(W_n + W_{j1} + \hbar\omega\right) = \left(W_n + \hbar\omega\right)^2 - W_{j1}^2 =$$
$$= W_n^2 + 2W_n\hbar\omega + \left(\hbar\omega\right)^2 - p_{j1}^2c^2 - W_n^2 = 2W_n\hbar\omega + \left[\left(\frac{\hbar\omega}{c}\right)^2 - p_{j1}^2\right]c^2 = 2W_n\hbar\omega,$$

where it is taken into account that the impulse of the molecule increases in size $p_{j1} = k = \frac{\hbar\omega}{c}$, and also $W_{j1}{}^2 = p_{j1}^2c^2 + W_n^2$. We believe that initially the molecule was in a state of rest.

Let's similarly transform the second addend, having multiplied its numerator and denominator on $W_n + W_{j11} - \hbar\omega^{/}$. Transformations to the denominator look like:

$$\left(W_n - W_{j11} - \hbar\omega^{/}\right)\left(W_n + W_{j11} - \hbar\omega^{/}\right) = \left(W_n - \hbar\omega^{/}\right)^2 - W_{j11}^2 =$$
$$= W_n^2 - 2W_n\hbar\omega^{/} + \left(\hbar\omega^{/}\right)^2 - p_{j11}^2c^2 - W_n^2 = -2W_n\hbar\omega^{/} + \left[\left(\frac{\hbar\omega^{/}}{c}\right)^2 - p_{j11}^2\right]c^2 = -2W_n\hbar\omega^{/},$$

where the decrease of the impulse of the molecule at radiation of the quantum equals $-p_{j11} = k^{/} = \frac{\hbar\omega^{/}}{c}$.

Thus, the formula (6.4.36) will be transformed to a form:

$$a_{kn}^{(1)}(t)=-\frac{1}{2W_n}\sum_s\sum_j\left[\begin{array}{l}\dfrac{V_{skj}^{/}V_{sjn}e^{\frac{i}{\hbar}(W_{j1}-W_n+sE)t}(W_n+W_{j1}+\hbar\omega)}{\hbar\omega(W_{j1}-W_n+sE)}-\\-\dfrac{V_{skj}V_{sjn}^{/}e^{\frac{i}{\hbar}(W_k-W_{j11}+sE)t}(W_n+W_{j11}-\hbar\omega^{/})}{\hbar\omega^{/}(W_k-W_{j11}+sE)}\end{array}\right].$$

In connection with the fact that in both variants of the quantum reradiated light of the same energy is absorbed, between intermediate energy and there is a connection $W_{j1}-W_n=W_k-W_{j11}$, figure 6.4.1. Hence, the previous expression can be transformed:

$$a_{kn}^{(1)}(t)=-\frac{1}{2W_n}\sum_s\sum_j\left[\begin{array}{l}\dfrac{V_{skj}^{/}V_{sjn}e^{\frac{i}{\hbar}(W_{j1}-W_n+sE)t}(W_n+W_{j1}+\hbar\omega)}{\hbar\omega(W_{j1}-W_n+sE)}-\\-\dfrac{V_{skj}V_{sjn}^{/}e^{\frac{i}{\hbar}(W_{j1}-W_n+sE)t}(W_n+W_{j11}-\hbar\omega^{/})}{\hbar\omega^{/}(W_{j1}-W_n+sE)}\end{array}\right]. \qquad (6.4.37)$$

Using Dirac's equation in Coulomb's to calibrate $\mathrm{div}\boldsymbol{A}=0$, Hamilton's operator looks like:

$$\hat{H}=c\boldsymbol{\alpha}\left(\hat{\boldsymbol{p}}-\frac{e}{c}\hat{\boldsymbol{A}}\right)+W_n\beta=c\boldsymbol{\alpha}\hat{\boldsymbol{p}}-e\boldsymbol{\alpha}\hat{\boldsymbol{A}}+e\phi+W_n\beta, \qquad (6.4.38)$$

where $\boldsymbol{A}$ is the vector-potential of a photon, ϕ - its scalar potential, - operator of the molecule impulse, e - a charge of the electron, - a matrix vector (a vector whose coordinate projections are square matrixes of 4th order, with components 0, ±1, ±i), is a square matrix of 4th order with components 0, ±1 [4]. All factors of matrix vector $\boldsymbol{\alpha}$ are four-vectors, of which the fourth component is imaginary [4]. Energy of the interaction of the photon-molecule characterizes only the second addend (6.4.38).

After absorption of a photon by a molecule the photon disappears. Therefore, there is the second addend in (6.4.38) to become equal to zero.

In states and a molecule is assumed to be a free particle on which the electromagnetic field does not act; therefore, the amplitude of its wave function satisfies Dirac's equation [4], where $\boldsymbol{r}$ is the radius-vector of the center of the mass of the molecule. Thus:

$$W_{j1}u_{j1} = \hat{H}_{j1}u_{j1} \text{ and } W_{j11}u_{j11} = \hat{H}_{j11}u_{j11}. \tag{6.4.39}$$

Operators of Hamilton are expressed by the formula (6.4.38) at $\boldsymbol{A} = \phi = 0$, since in states and there are no photons:

$$\hat{H}_{j1} = c\boldsymbol{\alpha}\hat{\boldsymbol{p}}_{j1} + W_n\beta = c\boldsymbol{\alpha}\hat{\boldsymbol{k}} + W_n\beta, \tag{6.4.40}$$

$$\hat{H}_{j11} = c\boldsymbol{\alpha}\hat{\boldsymbol{p}}_{j11} + W_n\beta = c\boldsymbol{\alpha}\hat{\boldsymbol{k}}^{/} + W_n\beta, \tag{6.4.41}$$

where impulses in the intermediate states and $\boldsymbol{p}_{j11} = -\boldsymbol{k}^{/}$ are taken into account. Components of a matrix $\beta = \pm 1$. Hence, it is possible to put and $W_{j11} - \beta W_n = -c\boldsymbol{\alpha}\boldsymbol{k}^{/}$. We accept $\beta = -1$, since initial and intermediate energy in additional factors and are summarized. Inserting these values in (6.4.37), we shall find:

$$a_{kn}^{(1)}(t) = -\frac{1}{2W_n}\sum_s\sum_j \left[\frac{V_{skj}^{/}V_{sjn}e^{\frac{i}{\hbar}(W_{j1}-W_n+sE)t}\left(c\boldsymbol{\alpha}\boldsymbol{k} + 1\cdot\hbar\omega\right)}{\hbar\omega\left(W_{j1} - W_n + sE\right)} + \frac{V_{skj}V_{sjn}^{/}e^{\frac{i}{\hbar}(W_{j1}-W_n+sE)t}\left(c\boldsymbol{\alpha}\boldsymbol{k}^{/} + 1\cdot\hbar\omega^{/}\right)}{\hbar\omega^{/}\left(W_{j1} - W_n + sE\right)} \right] =$$

$$= -\frac{1}{2W_n}\sum_s\sum_j \left[\left(V_{skj}^{/}V_{sjn}\left(\frac{\boldsymbol{\alpha}\boldsymbol{k}}{k} + 1\right) + V_{skj}V_{sjn}^{/}\left(\frac{\boldsymbol{\alpha}\boldsymbol{k}^{/}}{k^{/}} + 1\right)\right) \frac{e^{\frac{i}{\hbar}(W_{j1}-W_n+sE)t}}{\left(W_{j1} - W_n + sE\right)} \right] \tag{6.4.42}$$

where modules of the photons' impulses before and after interaction with a molecule equal and $k^{/} = \frac{\hbar\omega^{/}}{c}$, a matrix 1 is the unit matrix of the fourth order.

Let's write the transformations:

$$a_{kn}^{(1)}(t) = -\frac{1}{2W_n}\sum_s\sum_j \left[\begin{pmatrix} V_{skj}^{/}V_{sjn}\left(\frac{\boldsymbol{\alpha}\boldsymbol{k}}{k} + 1\right) + \\ + V_{skj}V_{sjn}^{/}\left(\frac{\boldsymbol{\alpha}\boldsymbol{k}^{/}}{k^{/}} + 1\right) \end{pmatrix} \frac{e^{\frac{i}{\hbar}(W_{j1}-W_n+sE)t}\left(\left(W_{j1} - W_n\right) - sE\right)}{\left(\left(W_{j1} - W_n\right)^2 - E^2\right)} \right] \tag{6.4.43}$$

As matrix elements of radiation we shall accept differences of corresponding energy transitions. In this case they equal $V_{skj}^{/} = W_{j1} - W_k = W_n - W_{j11} = V_{sjn}^{/}$. Hence, it is necessary to assume that thematrix elements of absorption are identical $V_{skj} = V_{sjn}$. Thus:

$$a_{kn}^{(1)}(t) = -\frac{1}{2W_n}\left(2 + \frac{\boldsymbol{ak}}{k} + \frac{\boldsymbol{ak}^{/}}{k^{/}}\right)\sum_s\sum_j\left[\frac{W_{jk}V_{sjn}e^{\frac{i}{\hbar}(W_{jn}+sE)t}\left(W_{jn} - sE\right)}{\left(W_{jn}^2 - E^2\right)}\right], \tag{6.4.44}$$

where it is designated and $W_{jk} = W_{j1} - W_k$. Addend 2 in the first bracket designates the sum of two unit matrixes.

Let's assume that the reaction of a molecule to electromagnetic influence by analogy with (6.4.19) looks like:

$$V_{sjn}e^{\frac{i}{\hbar}(W_{jn}+sE)t}\left(W_{jn} - sE\right)\frac{W_{jk}}{W_n} = -\left(\boldsymbol{p}_{jn}\boldsymbol{E} + \boldsymbol{m}_{jn}\boldsymbol{H}\right)\left(W_{jn} + E\right), \tag{6.4.45}$$

where is the ratio of energy change of a molecule at radiation of quantum to the energy of a molecule in the stationary state characterizes the reaction of radiation [21].

Hence, the formula (6.4.44) will be transformed:

$$a_{kn}^{(1)}(t) = \left(2 + \frac{\boldsymbol{ak}}{k} + \frac{\boldsymbol{ak}^{/}}{k^{/}}\right)\sum_j\left[\frac{\left(\boldsymbol{p}_{jn}\boldsymbol{E} + \boldsymbol{m}_{jn}\boldsymbol{H}\right)\left(W_{jn} + E\right)}{\left(W_{jn}^2 - E^2\right)}\right]. \tag{6.4.46}$$

In (6.4.46) a summation on two spin states is executed.

Wave function of the process of interaction (6.4.9) looks like:

$$\Psi_n^{(1)} = \sum_k \Psi_k^{(0)}\left(2 + \frac{\boldsymbol{ak}}{k} + \frac{\boldsymbol{ak}^{/}}{k^{/}}\right)\sum_j\left[\frac{\left(\boldsymbol{p}_{jn}\boldsymbol{E} + \boldsymbol{m}_{jn}\boldsymbol{H}\right)\left(W_{jn} + E\right)}{\left(W_{jn}^2 - E^2\right)}\right]. \tag{6.4.47}$$

To find the values of induced electric and magnetic dipole moments of molecule in a quantum state *n,* we use formulas (6.4.16).

As a first approximation the value of the electric dipole moment is:

$$\boldsymbol{p}_n(t)=\boldsymbol{p}_n^{(1)}(t)=\sum_k \mathrm{R}\int_q \Psi_k^{(0)*}\hat{\boldsymbol{p}}\Psi_n^{(1)}dq=$$
$$=\left(2+\frac{\alpha \boldsymbol{k}}{k}+\frac{\alpha \boldsymbol{k}^{/}}{k^{/}}\right)\sum_k \mathrm{R}\int_q \Psi_k^{(0)*}\hat{\boldsymbol{p}}\Psi_k^{(0)}dq\sum_j\left[\frac{(\boldsymbol{p}_{jn}\boldsymbol{E}+\boldsymbol{m}_{jn}\boldsymbol{H})(W_{jn}+E)}{(W_{jn}^2-E^2)}\right]=$$
$$=\left(2+\frac{\alpha \boldsymbol{k}}{k}+\frac{\alpha \boldsymbol{k}^{/}}{k^{/}}\right)\sum_k \mathrm{R}\boldsymbol{p}_{kn}\sum_j\left[\frac{(\boldsymbol{p}_{jn}\boldsymbol{E}+\boldsymbol{m}_{jn}\boldsymbol{H})(W_{jn}+E)}{(W_{jn}^2-E^2)}\right] \quad (6.4.48)$$

The value of the magnetic dipole moment is:

$$\boldsymbol{m}_n(t)=\left(2+\frac{\alpha \boldsymbol{k}}{k}+\frac{\alpha \boldsymbol{k}^{/}}{k^{/}}\right)\sum_k \mathrm{R}\boldsymbol{m}_{kn}\sum_j\left[\frac{(\boldsymbol{p}_{jn}\boldsymbol{E}+\boldsymbol{m}_{jn}\boldsymbol{H})(W_{jn}+E)}{(W_{jn}^2-E^2)}\right] \quad (6.4.49)$$

Furthermore, using the harmonious character of influence on a molecule, we have a formula:

$$(\boldsymbol{p}_{jn}\boldsymbol{E}+\boldsymbol{m}_{jn}\boldsymbol{H})(W_{jn}+E)=\left(\boldsymbol{p}_{jn}\boldsymbol{E}W_{jn}+i\hbar\boldsymbol{p}_{jn}\frac{\partial \boldsymbol{E}}{\partial t}+\boldsymbol{m}_{jn}\boldsymbol{H}W_{jn}+i\hbar\boldsymbol{m}_{jn}\frac{\partial \boldsymbol{H}}{\partial t}\right) \quad (6.4.50)$$

Let's substitute (6.4.50) in (6.4.48) and (6.4.49). As well as in paragraph 6.4.1 it is neglected by the quadrupole sizes proportional owing to their small size. As a result, we shall find:

$$\boldsymbol{p}_n=\frac{1}{3}\left(2+\frac{\alpha \boldsymbol{k}}{k}+\frac{\alpha \boldsymbol{k}^{/}}{k^{/}}\right)\sum_j\frac{1}{W_{jn}^2-E^2}\begin{pmatrix}(\boldsymbol{p}_{jn}\boldsymbol{p}_{jn})\boldsymbol{E}W_{jn}+\\ +\mathrm{R}(\boldsymbol{p}_{jn}\boldsymbol{m}_{jn})\boldsymbol{H}W_{jn}-\hbar\,\mathrm{Im}(\boldsymbol{p}_{jn}\boldsymbol{m}_{jn})\frac{\partial \boldsymbol{H}}{\partial t}\end{pmatrix} \quad (6.4.51)$$

$$\boldsymbol{m}_n=\frac{1}{3}\left(2+\frac{\alpha \boldsymbol{k}}{k}+\frac{\alpha \boldsymbol{k}^{/}}{k^{/}}\right)\sum_j\frac{1}{W_{jn}^2-E^2}\left(\mathrm{R}(\boldsymbol{p}_{jn}\boldsymbol{m}_{jn})\boldsymbol{E}W_{jn}+\hbar\,\mathrm{Im}(\boldsymbol{p}_{jn}\boldsymbol{m}_{jn})\frac{\partial \boldsymbol{E}}{\partial t}\right) \quad (6.4.52)$$

If we are to assume the reaction of a molecule to electromagnetic influence as (6.4.26), the factors look like:

$$\alpha=\frac{1}{3\hbar}\left(2+\frac{\alpha \boldsymbol{k}}{k}+\frac{\alpha \boldsymbol{k}^{/}}{k^{/}}\right)\sum_j\frac{\omega_{jn}\boldsymbol{p}_{jn}\boldsymbol{p}_{jn}}{\omega_{jn}^2-\omega^2}, \quad (6.4.53)$$

$$\beta = \frac{c}{3\hbar}\left(2 + \frac{\boldsymbol{\alpha k}}{k} + \frac{\boldsymbol{\alpha k}'}{k'}\right)\sum_j \frac{\mathrm{Im}\left(\boldsymbol{p}_{jn}\boldsymbol{m}_{jn}\right)}{\omega_{jn}^2 - \omega^2}, \tag{6.4.54}$$

$$\gamma = \frac{1}{3\hbar}\left(2 + \frac{\boldsymbol{\alpha k}}{k} + \frac{\boldsymbol{\alpha k}'}{k'}\right)\sum_j \frac{\omega_{jn}\mathrm{R}\left(\boldsymbol{p}_{jn}\boldsymbol{m}_{jn}\right)}{\omega_{jn}^2 - \omega^2}. \tag{6.4.55}$$

In the formulas (6.4.53) - (6.4.55) is the frequency of resonance at the interaction of a molecule with radiation on electronic transition $n \rightarrow j$. We shall note that, as opposed to (6.4.27) - (6.4.29) in this case, factors have matrix character, since 2 is the sum of two unit matrixes, and $\boldsymbol{\alpha}$ - a matrix vector.

We consider how formulas (6.4.53) - (6.4.55) correspond with formulas for a non-relativistic case (6.4.27) - (6.4.29). The double unit matrix determines the first addend in factor and the matrixes of matrix vector $\boldsymbol{\alpha}$ determine the second and the third addends of the even order. Matrixes of the even order have no common component in mutually perpendicular diagonals. Moreover, non-zero components of a unit matrix always have a corresponding zero component of matrix vector $\boldsymbol{\alpha}$. Therefore, if a non-zero diagonal component of an unit matrix is used, the addends with a matrix vector are equal to zero and $\left(2 + \frac{\boldsymbol{\alpha k}}{k} + \frac{\boldsymbol{\alpha k}'}{k'}\right) = (2) = 2$. If zero is used in non-diagonal components of the unit matrix of the first addend, at the direct reradiated quantum or zero.

Let's consider the direct reradiated quantum on a molecule in a direction of an axis *x*. Such a situation is observed, for example, in the research into the optical activity of substances. In this case $\left(\frac{\alpha_x k_x}{k_x} + \frac{\alpha_x k_x'}{k_x'}\right) = (2\alpha_x)$. Taking into account the *x*-projection of a matrix vector [4], it is possible to notice that for unit components of this vector a bracket $\left(2 + \frac{\boldsymbol{\alpha k}}{k} + \frac{\boldsymbol{\alpha k}'}{k'}\right) = 2$. If the components of the unit matrix and matrix vector are simultaneously equal to zero, the optical activity of a molecule is absent.

For direct reradiated quantum on a molecule it is possible to enter a matrix $\tau_{11} = (1 + \alpha_x) = \begin{pmatrix} 1 & 0 & 0 & 1 \\ 0 & 1 & 1 & 0 \\ 0 & 1 & 1 & 0 \\ 1 & 0 & 0 & 1 \end{pmatrix}$. In this case, formulas (6.4.53) - (6.4.55) can be copied as being practically identical to the non-relativistic case (6.4.27) - (6.4.29):

$$\alpha = \frac{2}{3\hbar}\tau_{11}\sum_j \frac{\omega_{jn}\boldsymbol{p}_{jn}\boldsymbol{p}_{jn}}{\omega_{jn}^2 - \omega^2}, \tag{6.4.56}$$

$$\beta = \frac{2c}{3\hbar}\tau_{11}\sum_j \frac{\mathrm{Im}(\boldsymbol{p}_{jn}\boldsymbol{m}_{jn})}{\omega_{jn}^2 - \omega^2}, \tag{6.4.57}$$

$$\gamma = \frac{2}{3\hbar}\tau_{11}\sum_j \frac{\omega_{jn}\mathrm{R}(\boldsymbol{p}_{jn}\boldsymbol{m}_{jn})}{\omega_{jn}^2 - \omega^2}. \tag{6.4.58}$$

As opposed to the non-relativistic variant, the formulas (6.4.56) - (6.4.58) take into account both an opportunity for the existence of optical activity and an opportunity for its absence at the occurrence of dipole the electric and magnetic moments of a molecule. I.e. in these formulas the opportunity for the presence of a connection of optical activity and of a kind of conformative structure of a molecule are incorporated: the existence of symmetry or its absence. In the absence of the symmetry of a molecule (first of all, the center of symmetry [8]) the molecule is optically active; i.e. it is chiral [52]. Chiral molecules appear in two forms: right and left (as with the right and left hands). These two configurations cannot be combined with each other in any rotation of a molecule in space. If a molecule has chirality, the environment is optically active; also, it is necessary to use a component of a matrix $\tau_{11} = 1$. Otherwise, it is necessary to use to a component $\tau_{11} = 0$.

Comparison of non-relativistic and relativistic analysis of the phenomenon of optical activity shows that the reaction of a molecule to electromagnetic influence in these two cases virtually does not differ from each other. In a relativistic variant there is the additional matrix factor symmetrical on two diagonals where components can have a size of 0 or 1.

This factor reflects the conformative structure of a molecule. For chiral molecules it is necessary to use matrix component 1; otherwise, a component 0.

In practical calculations it is convenient to use the following formula for rotary force [63]:

$$R_{kn} = \frac{3hc}{8\pi^3 n_0}\int_0^\infty \frac{\theta_{kn}(\lambda)}{\lambda}d\lambda. \tag{6.4.59}$$

where is an elliptic of light (function of wave length λ) interacting with the transition; $n \to k$, is the numerical concentration of molecules.

It is convenient to use the resulting rotary force which is numerically equal in value, commensurable with unit:

$$R_{kn}^{/} = \frac{100}{\mu_Б \mu_D} R_{kn}, \tag{6.4.60}$$

where it is the Bohr magneton; is equal to one Debye - a unit of measurements of the electric dipole moment. If we were to carry out the calculation of the resulting rotary force of one mol in system CGSE it is possible to use the formula:

$$R_{kn}^{/} = 0.75 \cdot 10^{-2} \int_{0}^{\infty} \frac{\theta_{kn}(\lambda)}{\lambda} d\lambda . \qquad (6.4.61)$$

6.4.3. Model of Coupled Oscillators

Practical application of the dependences for the rotary ability of environment β (6.4.57) and for rotary force (6.4.48) is impossible without the use of a concrete molecular model.

Let's consider one such model referred to – a model of coupled oscillators. Such a model has been investigated, for example, in [54, 64]. The basic requirement to model it is asymmetry.

Let's say that in a molecule there are two coupled particles (atoms) 1 and 2, with coordinates and $\left(0, Y_2, +\frac{d}{2}\right)$, where $d = const$, figure 6.4.2

The potential energy of the system of particles – in the system there are several degrees of freedom - looks like the square-law form:

$$U = \frac{1}{2} \sum_{i,j} k_{ij} X_i X_j , \qquad (6.4.62)$$

where is the constant factor of elastic constraint of particles with the centers of oscillations and with each other. Thus, there is [55].

Applying the formula (6.4.62) for the examined system of 2 particles, for each of which there are elastic constraints with the position of balance and with other particle, we shall find:

$$U = \frac{1}{2} k_{11} X_1^2 + k_{12} X_1 Y_2 + \frac{1}{2} k_{22} Y_2^2 . \qquad (6.4.63)$$

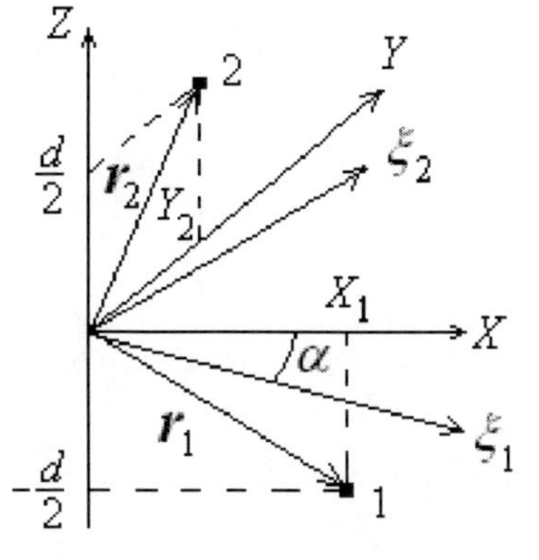

Figure 6.4.2.

We accept that oscillations are possible for the particle 1 only on axis *X*, and for the particle 2 on axis *Y*.

The kinetic energy of a system of two particles looks like:

$$T = \frac{1}{2} m_1 V_1^2 + \frac{1}{2} m_2 V_2^2, \tag{6.4.64}$$

where and are speeds of oscillating particles, and - its mass.

Let's enter a new system of coordinates and ξ_2 , which are rotated about axis *Z* on angle α, see figure 6.4.2.

The connection between coordinates X_1, and ξ_1 , looks like:

$$X_1 \sqrt{m_1} = \xi_1 \cos\alpha + \xi_2 \sin\alpha ,$$
$$Y_2 \sqrt{m_2} = -\xi_1 \sin\alpha + \xi_2 \cos\alpha . \tag{6.4.65}$$

Substituting (6.4.65) in (6.4.64) we find the kinetic energy of system as:

$$T = \frac{1}{2}\left(V_{\xi 1}^2 + V_{\xi 2}^2\right), \tag{6.4.66}$$

where and $V_{\xi 2} = \dfrac{\partial \xi_2}{\partial t}$.

Substituting (6.4.65) in (6.4.63) we shall find the potential energy of the system as:

$$U = \frac{1}{2}\left[(\omega_1 \xi_1)^2 + (\omega_2 \xi_2)^2\right], \tag{6.4.67}$$

where squares of their own oscillating frequencies of particles 1 and 2 are designated:

$$\omega_1^2 = \frac{k_{11}}{m_1}\cos^2\alpha - \frac{2k_{12}}{\sqrt{m_1 m_2}}\sin\alpha\cos\alpha + \frac{k_{22}}{m_2}\sin^2\alpha \cdot \tag{6.4.68}$$

$$\omega_2^2 = \frac{k_{11}}{m_1}\sin^2\alpha + \frac{2k_{12}}{\sqrt{m_1 m_2}}\sin\alpha\cos\alpha + \frac{k_{22}}{m_2}\cos^2\alpha \cdot \tag{6.4.69}$$

Besides the condition for angle α is accepted as:

$$\left(\frac{k_1}{m_1}-\frac{k_2}{m_2}\right)\sin 2\alpha+\frac{k_{12}}{\sqrt{m_1 m_2}}\cos 2\alpha=0. \tag{6.4.70}$$

Thus, in coordinates and under the condition of (6.4.70), the oscillaions of particles 1 and 2 are quasi-independent with cyclic frequencies; accordingly, and ω_2. The condition (6.4.70) defines an angle of rotation of coordinates system about axis Z, figure 6.4.2.

Let's consider the influence on the system of particles 1 and 2 electromagnetic waves in case particles (ions) have charges and e_2.

The electric dipole moment of system equals:

$$\boldsymbol{p}=\sum e_i \boldsymbol{r}_i=e_1 X_1 \boldsymbol{i}+e_2 Y_2 \boldsymbol{j}, \tag{6.4.71}$$

where $\boldsymbol{i}$, $\boldsymbol{j}$, are the unit vectors of coordinates.

The magnetic moment of the system we shall find under the formula [21]:

$$\boldsymbol{m}=\frac{1}{2c}\sum e_i \boldsymbol{r}_i\times \boldsymbol{V}_i=\frac{1}{2c}\left[e_1\boldsymbol{r}_1\times \boldsymbol{V}+e_2\boldsymbol{r}_2\times \boldsymbol{V}_2\right], \tag{6.4.72}$$

where and are radius vectors of particles 1 and 2, figure 6.4.2.

According to this figure:

$$\boldsymbol{r}_1=-\frac{d}{2}\boldsymbol{k}+X_1\boldsymbol{i};\ \boldsymbol{r}_2=\frac{d}{2}\boldsymbol{k}+Y_2\boldsymbol{j}, \tag{6.4.73}$$

Finding derivatives on time from (6.4.73), we have:

$$\boldsymbol{V}_1=V_1\boldsymbol{i};\ \boldsymbol{V}_2=V_2\boldsymbol{j}. \tag{6.4.74}$$

Hence, the magnetic moment of the particles system equals:

$$\boldsymbol{m}=-\frac{d}{2c}\left(e_1 V_1\boldsymbol{j}+e_2 V_2\boldsymbol{i}\right). \tag{6.4.75}$$

The classical value of rotary force of transition we shall find under the formula (6.4.30), using (6.4.71) and (6.4.75):

$$R_{kn}=\mathrm{Im}\left(\boldsymbol{p}_{kn}\boldsymbol{m}_{kn}\right)=\boldsymbol{p}\cdot\boldsymbol{m}=-\frac{d}{2c}e_1e_2\left(X_1V_2+Y_2V_1\right)=-\frac{d}{2c}e_1e_2\frac{\partial X_1Y_2}{\partial t}. \tag{6.4.76}$$

TheIm of allocation of the imaginary addend in the classical approach to the model of a role does not occur.

Having substituted in (6.4.76) the formulas (6.4.65), we shall transition to coordinates and ξ_2 :

$$R_{kn} = -\frac{d}{2c}\frac{e_1 e_2}{\sqrt{m_1 m_2}}\frac{\partial}{\partial t}\left[\left(\xi_2^2 - \xi_1^2\right)\sin\alpha\cos\alpha + \xi_1\xi_2\cos 2\alpha\right]. \quad (6.4.77)$$

In connection with the fact that oscillations of particles in coordinates and are quasi-independent, the last addend should not enter into the expression for rotary force. Therefore:

$$R_{kn} = -\frac{d}{2c}\frac{e_1 e_2}{\sqrt{m_1 m_2}}\frac{\partial}{\partial t}\left[\left(\xi_2^2 - \xi_1^2\right)\sin\alpha\cos\alpha\right] = -\frac{d}{c}\frac{e_1 e_2}{\sqrt{m_1 m_2}}\left(V_{\xi 2}\xi_2 - \xi_1 V_{\xi 1}\right)\sin\alpha\cos\alpha \quad (6.4.78)$$

We carry out the quantum transition in the model.

Matrix components of possible transitions are characterized by corresponding quantum numbers.

Quantum transitions on coordinates also can be found from the quantum theory of linear quantum oscillator [4]:

$$\langle n_1|\xi_1|n_1'\rangle = \sqrt{\frac{\bar{n}_1\hbar}{2m_1\omega_1}}\,; \quad (6.4.79)$$

Where, for coordinate ξ_1, the quantum numbers are $n_1' = n_1 \pm 1$, and quantum number is greater than the numbers and n_2. It is similar for coordinate ξ_2.

In total, 4 possible transitions from a condition with quantum numbers and n_2 are received:

$$(n_1+1, n_2),\ (n_1-1, n_2),\ (n_1, n_2+1),\ (n_1, n_2-1). \quad (6.4.80)$$

In the Heisenbergen representation of quantum mechanics in the formula (6.4.78) it is necessary to replace classical sizes with quantum-mechanical operators:

$$R_{kn} = \mathrm{Im}\left\{-\frac{d}{4c}\frac{e_1 e_2}{\sqrt{m_1 m_2}}\left(\hat{V}\hat{\xi} - \hat{\xi}\hat{V}\right)\sin\alpha\cos\alpha\right\}. \quad (6.4.81)$$

Two particles in a molecule are considered as a uniform quantum system.

The number four in a denominator for separate transition arises owing to the account (6.4.80).

Using a ratio of quantum operator commutation [4]:

$$\hat{V}\hat{\xi} - \hat{\xi}\hat{V} = -i\hbar, \tag{6.4.82}$$

we find the rotary force of transition $n \to k$:

$$R_{kn} = \frac{d}{4c}\frac{\hbar e_1 e_2}{\sqrt{m_1 m_2}}\sin\alpha\cos\alpha\,. \tag{6.4.83}$$

We shall find the rotating ability of the environment consisting from a pair of coupled oscillators for an electromagnetic wave of cyclic frequency ω under the formula (6.4.31):

$$\beta = \frac{2c}{3\hbar}\sum_k \frac{R_{kn}}{\omega_{kn}^2 - \omega^2} = \frac{d}{6}\frac{e_1 e_2}{\sqrt{m_1 m_2}}\sin\alpha\cos\alpha\left\{\frac{1}{\omega_1^2 - \omega^2} + \frac{1}{\omega_2^2 - \omega^2}\right\}, \tag{6.4.84}$$

where squares of own (resonant) frequencies of particles also can be found under formulas (6.4.68) and (6.4.69).

The angle of polarization plane rotation of the optically active substance can be calculated under the formula (2.4.42), with use (1.2.13) (in system CGS $\frac{\varepsilon - 1}{\varepsilon + 2} = \frac{4\pi}{3}\alpha n_0$):

$$\begin{gathered}\varphi = p^2\frac{(\varepsilon - 1)}{\alpha n}\beta l = p^2 n_0\frac{\varepsilon + 2}{3n}\beta l = \\ = \frac{2\pi n_0}{9n}(\varepsilon + 2)d\frac{e_1 e_2}{\sqrt{m_1 m_2}}\sin\alpha\cos\alpha\left\{\frac{1}{\omega_1^2 - \omega^2} + \frac{1}{\omega_2^2 - \omega^2}\right\}p^2 l\end{gathered} \tag{6.4.85}$$

The formula (6.4.85) for the first time has been deduced in [65].

Passing to wave lengths, in view of wave number $p = \frac{n\omega}{c}$, and also $\frac{\omega_i}{\omega} = \frac{\lambda}{\lambda_i}$, we shall find:

$$\varphi = \frac{2\pi n n_0}{9c^2}(\varepsilon + 2)d\frac{e_1 e_2}{\sqrt{m_1 m_2}}\sin\alpha\cos\alpha\left\{\frac{\lambda_1^2}{\lambda^2 - \lambda_1^2} + \frac{\lambda_2^2}{\lambda^2 - \lambda_2^2}\right\}l, \tag{6.4.86}$$

The formula (6.4.85) reflects the basic properties of the optically active substance. First, oscillators should be coupled, since if $k_{12} = 0$, according to (6.4.70) the angle $\alpha = 0$. Hence, the rotation of the polarization plane disappears $\varphi = 0$. Second, the oscillators (atoms, ions) should be from each other at some distance *d*. At $d = 0$, an angle of the polarization plane rotation of light $\varphi = 0$.

Usually for small molecules there is $\frac{d}{\lambda} \approx 10^{-3}$. The distance *d* results in a difference of phases of a light wave within the limits of one molecule. This difference of phases is a necessary condition of optical rotation existence. The necessary asymmetry of a molecule was mentioned previously regarding a choice of model.

6.4.4. Spectral Dependences of Optical Rotation

Research of spectra of optical rotation allows us to receive much information on a complex structure; for example, albuminous molecules. There is only a small probability that an albuminous molecule with its huge mass, usually some tens of thousands of units, has any symmetry. The spectrum of optical rotation allows the reception of many characteristics of organic molecules; in particular character, their asymmetry, a configuration of a molecule, conformative mobility, etc.

For example, for definition of the molecular configuration it is necessary to compare a curve of dispersion of optical rotation of the researched substance with a curve of dispersion of optical rotation of another substance in which the configuration of a molecule is known also which contains identical chromophore in the same conformative environment. Frequently the method of research of a dispersion of optical rotation appears to be the only thing at the decision of some chemical problems.

The basic part of all research into finding spectra of optical rotation, approximately 95%, is carried out by chemists/organicists. The majority of measurements of a dispersion of optical rotation lays in a range of wave lengths $\lambda = 450 - 700\ nm$, i.e. in a seen region.

Possible forms of spectral characteristics of circular double-refraction and circular dichroism are schematically shown in figure 6.4.3.

In the figure are shown: specific rotation α (2.4.2) and an elliptical θ (2.4.6). Curves *a* and *b* concern the positive (D) Cotton effect (circular dichroism), and *c* and *d* to negative (L) Cotton effect.

As a whole, the explanation of the Cotton effect consists in the fact that the electromagnetic field of a light ray influences the molecular orbital electrons. It results in the redistribution of electronic charge in the molecule. For small molecules the ratio of its sizes to the light wave length is $\sim 10^{-3}$; therefore, there is a difference of the light wave phase within the limits of one molecule. It is necessary to take into account also the interaction of molecular electrons. In the absence of such interaction as it was already marked, there is no optical activity either. Besides, the optically active molecules should not have a plane or center of symmetry.

In connection with the fact that the moving of the charges into the molecule cannot take place before the occurrence of electromagnetic field which causes this moving, molecular rotation (2.4.3) and molecular ellipticality (2.4.7) submit to Kronig-Kramers' ratio [56]:

$$M = \frac{2}{\pi} \int_0^{\infty} \theta_{mol}(\lambda') \frac{\lambda'}{\lambda^2 - \lambda'^2} d\lambda' , \tag{6.4.87}$$

$$\theta_{mol} = -\frac{2}{\pi\lambda} \int_0^{\infty} M(\lambda') \frac{\lambda'^2}{\lambda^2 - \lambda'^2} d\lambda' , \tag{6.4.88}$$

where λ there is wave length of light, - parameter (wave length) of integration.

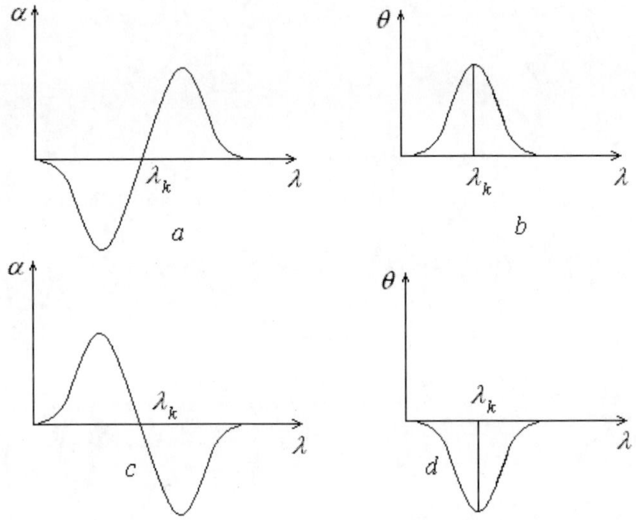

Figure 6.4.3.

Let's assume that elliptically, figure 6.4.3b, with the positive Cotton effect it is possible to present approximately as a Gaussonian curve:

$$\theta = \theta_{max} e^{-\frac{(\lambda - \lambda_k)^2}{\Delta_k^2}} , \tag{6.4.89}$$

where is the maximal value of ellipticality, - an interval of wave lengths at which θ falls up to of the maximal value, figure 6.4.4; this size refers to the half-width of the interaction band.

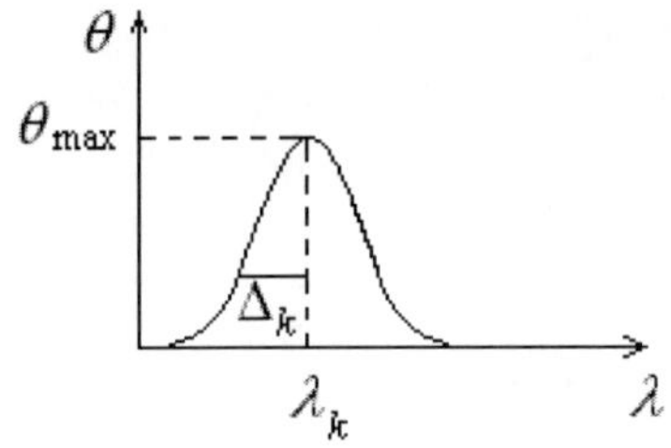

Figure 6.4.4.

Representation (6.4.89) is a consequence of expansion into the series

$$\frac{1}{e^{\frac{(\lambda-\lambda_k)^2}{\Delta_k^2}}} \approx \frac{1}{1+\frac{(\lambda-\lambda_k)^2}{\Delta_k^2}} = \frac{\Delta_k^2}{(\lambda-\lambda_k)^2+\Delta_k^2} \approx \frac{4\lambda^2\Delta_k^2}{(\lambda^2-\lambda_k^2)^2+4\lambda^2\Delta_k^2},$$

resulting in a standard resonant curve in area $\lambda \approx \lambda_k$.

Using (6.4.87) and passing to molecular ellipticality, we shall find the molecular rotation:

$$M = \frac{2}{\pi}\theta_{\max}\int_0^\infty e^{-\frac{(\lambda'-\lambda_k)^2}{\Delta_k^2}} \frac{\lambda'}{\lambda^2-\lambda'^2}d\lambda'. \qquad (6.4.90)$$

With some approximation we have:

$$M = \frac{2}{\sqrt{\pi}}\theta_{\max}\left[e^{-\frac{(\lambda-\lambda_k)^2}{\Delta_k^2}}\int_0^{\frac{(\lambda-\lambda_k)}{\Delta_k}} e^{-X^2}dX - \frac{\Delta_k}{2(\lambda+\lambda_k)}\right]. \qquad (6.4.91)$$

The received dependence of the molecular rotation on the wave length there looks similar to figure 6.4.3a. Let's find the rotary force of the electronic transition in a molecule under the formula (6.4.59), using (6.4.89):

$$R_k = \frac{3hc}{8\pi^3 n_0}\int_0^\infty \frac{\theta_k(\lambda)}{\lambda}d\lambda = \frac{3hc\theta_{\max}}{8\pi^3 n_0}\int_0^\infty \frac{e^{-\frac{(\lambda-\lambda_k)^2}{\Delta_k^2}}}{\lambda}d\lambda = \gamma\sqrt{\pi}\theta_{\max}\frac{\Delta_k}{\lambda_k}, \qquad (6.4.92)$$

where γ is the constant factor dependent on the used system of units of measurements.

Excluding from the system (6.4.91) and (6.4.92) size θ_{max}, we obtain:

$$M = \frac{2}{\pi} R_k \frac{\lambda_k}{\gamma\Delta_k}\left[e^{-\frac{(\lambda-\lambda_k)^2}{\Delta_k^2}} \int_0^{\frac{(\lambda-\lambda_k)}{\Delta_k}} e^{-X^2} dX - \frac{\Delta_k}{2(\lambda+\lambda_k)} \right]. \tag{6.4.93}$$

Thus, molecular rotation is the proportionally rotating force.

As the size it tends to $\pm\infty$, the function is an asymptotical approach to $\frac{\Delta_k}{2(\lambda-\lambda_k)}$.

Hence, in the areas far from the center of the interaction band where $\lambda - \lambda_k >> \Delta_k$, the formula (6.4.93) gets a kind:

$$M = \frac{2}{\pi} R_k \frac{\lambda_k}{\gamma\Delta_k}\left[\frac{\Delta_k}{2(\lambda-\lambda_k)} - \frac{\Delta_k}{2(\lambda+\lambda_k)} \right] = \frac{2}{\pi\gamma} R_k \frac{\lambda_k^2}{\lambda^2 - \lambda_k^2}. \tag{6.4.94}$$

The received formula expresses Drude's law which is correct for the areas concerning far from the center of interaction band (resonance). Molecular rotation does not depend from half-width of interaction bands Δ_k.

Taking into account the fact that half-width of interaction bands defines circular dichroism, it is possible to conclude that optical rotation and circular dichroism are rather independent physical phenomena.

6.5. Occurrence of Brake Electromagnetic Radiation at an Influence on Substance by the β-Particles

The influence of ionizing radiations on a substance is carried out due to various physical phenomena. For example, under the influence of X-ray and γ-radiations on a substance the processes of coherent and non-coherent scattering, an internal photoeffect, forming a pair of electron - positron, etc. are observed.

Influence α-, β-, proton, neutron, etc. radioactive radiations are also accompanied by various phenomena.

For example, we shall consider some physical processes at the interaction of a flux of β-particles, i.e. electrons with substance.

The following phenomena are possible at the interaction of β-particles with substance [57]:

- elastic collisions of electrons with atoms at which the energy of electrons practically does not vary;
- non-elastic collisions of electrons with atoms at which the energy of electrons is lost on the excitation and ionization of atoms;
- non-elastic collisions of electrons with the nucleus and electrons of the atoms, accompanied by formation of brake radiation.

The energy of electrons of β - rays depends on the radioactive element which radiates these rays. For a spectrum β - rays typically exist on the top power border. For example, isotope has two spectra β - rays. The top border of a low-energy spectrum of electrons is $E_0 = 318\ keV$, and high energy spectrum [58].

The basic fraction of β - rays – 99.9% falls at a low-energy spectrum. Short-wave boundary length of a wave of brake radiation (if all energy of the electron is transferred to a photon) for a low-energy spectrum is $\lambda_{\min} = \dfrac{2\pi c\hbar}{E_0} = 0.0039\, nm$, for a high-energy spectrum is $\lambda_{\min} = \dfrac{2\pi c\hbar}{E_0} = 0.00083\, nm$, where c is a speed of light in vacuum. The average lengths of waves calculated on the average energy of electrons are correspondingly and $\lambda_a = 0.00032\ nm$. All of these lengths of a wave lay in the region of short-wave X-ray.

Electrons of β - rays will rather deeply penetrate into the substance; for example, tissues of an organism.

Electrons with energy $150\, keV$ will penetrate at a depth of $0.278\, mm$, and with energy at a depth of $19\, cm$. For electrons the linear energy transfer (LET) sharply falls with an increase of their energy and makes for energy above [59].

6.5.1. Analysis of the Occurrence of Brake Radiation at Movement of the Electron in a Substance. Non-Relativistic Variant

Let's consider in more detail braking with scattering the electron moving in the field of a nucleus, figure 6.5.1.

During flight of the electron about a nucleus there is a dipole configuration of charges. The electric dipole is directed usually from an electron to the nucleus. For convenience, we shall consider that it is directed against a unit vector ***n*** which is directed on a radius-vector of the electron. During the movement of the electron the dipole moment rotates in a plane of scattering of the electron. The maximal probability of a start of quantums of brake radiation is in the direction of perpendicular axes of a dipole. This probability is distributed in spherical coordinates under the sine law. The diagram of radiation of quantums is shown in figure 6.5.1.

Assuming that at the interaction of moving one electron with an electric field of a nucleus one photon is radiated, we shall write down the operator (Hamiltonian) radiations of a photon with an impulse ***k*** by system electron - nucleus [4]:

$$\hat{H} = -\frac{e}{mc}\hat{\boldsymbol{p}}\boldsymbol{A}, \tag{6.5.1}$$

where *e* is a charge of the electron, *m* - its mass, *Z* - charging number of a nucleus, - a radius-vector up to a considered point of space in a direction of radiation of a photon, - the operator of an impulse of the electron [53], - the vector-potential of a photon is radiating periodic function of coordinates (the radius *r* is directed at an impulse of a photon *k*, which is why used scalar forms a radius and an impulse), ω - cyclic frequency of a photon, - a unit vector of photon polarization which is always perpendicular to a vector of an impulse of a photon $\boldsymbol{k}$, *V* - some normalizing volume, - classical radius of the electron in system CGS, see (3.1.6).

It is taken into account that the formula for vector-potential should not include the mass of the electron, but in it there should be a value of potential energy of electric field $U = \frac{ee_1}{r}$, where is a charge of nucleus about which the photon is propagated. The factor in the formula for vector-potential is the ratio of potential energy of a field to energy of the radiated photon. All transformations are carried out in the CGS system as it is accepted in quantum mechanics.

Let's find a matrix element of the interaction energy of an electron moving in an electric field of a nucleus, at a change of an impulse of the electron $\boldsymbol{p} \rightarrow \boldsymbol{p}'$, where $\boldsymbol{p}$ is an initial impulse of the electron, and - its impulse after scattering on a nucleus. We assume a nucleus is the heavy motionless center. In this case the change of an impulse of the electron on the module is equal to an impulse of the radiated photon $|\boldsymbol{q}| = |\boldsymbol{k}|$.

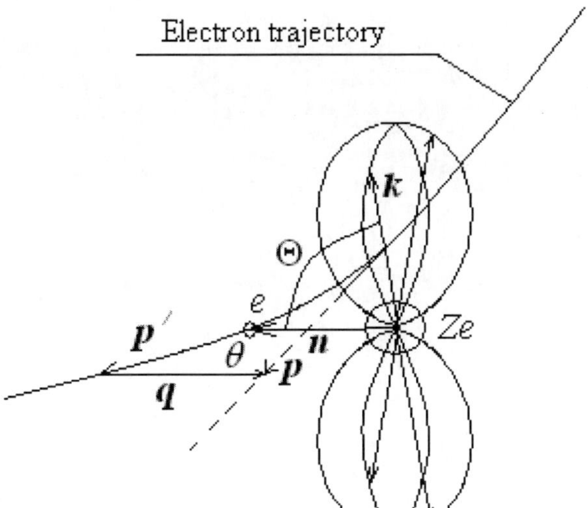

Figure 6.5.1.

In vector-potential $\boldsymbol{A}$ it is necessary to transition from a radius-vector in the direction of a photon radiation to a radius-vector in the direction of the electron since further wave functions of the electron with its radius-vector will be used. Taking into account an angle $\angle\Theta = \angle\boldsymbol{k},\boldsymbol{n}$, fig. 6.5.1, we shall write down:

$$\boldsymbol{A}(\boldsymbol{r}) = \sqrt{\frac{2\pi\hbar c^2}{\omega V}}\boldsymbol{e}\frac{mcZr_0}{kr}e^{-\frac{i}{\hbar}\boldsymbol{kr}} = \sqrt{\frac{2\pi\hbar c^2}{\omega V}}\boldsymbol{e}\frac{mcZr_0}{kr}e^{-\frac{i}{\hbar}|\boldsymbol{k}|\cdot|\boldsymbol{r}|\cos(\boldsymbol{k},\boldsymbol{n})}$$

$$= \sqrt{\frac{2\pi\hbar c^2}{\omega V}}\boldsymbol{e}\frac{mcZr_0}{qr}e^{-\frac{i}{\hbar}|\boldsymbol{q}|\cdot|\boldsymbol{r}|\cos\Theta} = \sqrt{\frac{2\pi\hbar c^2}{\omega V}}\boldsymbol{e}\frac{mcZr_0}{qr}e^{-\frac{i}{\hbar}\boldsymbol{qr}}$$

Simultaneously we use replacement $|\boldsymbol{q}| = |\boldsymbol{k}|$. A unit vector in a direction of the electron is $\boldsymbol{n} = \frac{\boldsymbol{r}}{r}$. In this case the scalar product in exponent conditionally means $\boldsymbol{qr} = qr\cos\Theta$, in spite of the fact that between vectors $\boldsymbol{q}$ and $\boldsymbol{r}$ the angle is equal to 180^0, figure 6.5.1.

In spite of the fact that transformations will be written at once concerning a radius-vector of the electron, the actual transition for convenience will be made later about what we shall discuss below.

The matrix element of energy of a photon radiation at the interaction of the electron and a field of a nucleus can be written down as (in this case $\boldsymbol{r}$ - a radius-vector of the electron):

$$H_{p'p} = \int_V \varphi^*_{p'}\hat{H}\varphi_p dV =$$

$$= -\frac{e}{mc}\int_V \varphi^*_{p'}\hat{\boldsymbol{p}}\boldsymbol{A}\varphi_p dV = -\frac{e}{mc}\int_V \varphi^*_{p'}\hat{\boldsymbol{p}}\boldsymbol{A}\varphi_p dV =$$

$$= -Ze\sqrt{\frac{2\pi\hbar c^2}{\omega V}}\int_V u'^* e^{-\frac{i}{\hbar}\boldsymbol{p}'\boldsymbol{r}}\left(\hat{\boldsymbol{p}}\boldsymbol{e}\frac{r_0}{qr}e^{-\frac{i}{\hbar}\boldsymbol{qr}}\right)ue^{\frac{i}{\hbar}\boldsymbol{pr}}dV =$$

$$= -Ze\sqrt{\frac{2\pi\hbar c^2}{\omega V}}\int_V u'^* e^{-\frac{i}{\hbar}\boldsymbol{p}'\boldsymbol{r}}\left(\boldsymbol{e}\hat{\boldsymbol{p}}\frac{r_0}{qr}e^{-\frac{i}{\hbar}\boldsymbol{qr}}\right)ue^{\frac{i}{\hbar}\boldsymbol{pr}}dV =$$

$$= -Ze\sqrt{\frac{2\pi\hbar c^2}{\omega V}}\int_V u'^* e^{-\frac{i}{\hbar}\boldsymbol{p}'\boldsymbol{r}}\left(-i\hbar\boldsymbol{e}r_0\nabla\frac{1}{qr}e^{-\frac{i}{\hbar}\boldsymbol{qr}}\right)ue^{\frac{i}{\hbar}\boldsymbol{pr}}dV =$$

$$= -Ze\sqrt{\frac{2\pi\hbar c^2}{\omega V}}\int_V u'^* e^{-\frac{i}{\hbar}\boldsymbol{p}'\boldsymbol{r}}\boldsymbol{e}\frac{r_0}{q}\left(-i\hbar\frac{1}{r}\nabla e^{-\frac{i}{\hbar}\boldsymbol{qr}} - i\hbar e^{-\frac{i}{\hbar}\boldsymbol{qr}}\nabla\frac{1}{r}\right)ue^{\frac{i}{\hbar}\boldsymbol{pr}}dV =$$

$$= -Ze\sqrt{\frac{2\pi\hbar c^2}{\omega V}}\frac{r_0}{q}\int_V u'^* e^{-\frac{i}{\hbar}\boldsymbol{p}'\boldsymbol{r}}\boldsymbol{e}\left(-\frac{1}{r}q\cos\theta\frac{\boldsymbol{r}}{r}e^{-\frac{i}{\hbar}\boldsymbol{qr}} + e^{-\frac{i}{\hbar}\boldsymbol{qr}}\hat{\boldsymbol{p}}\left(\frac{1}{r}\right)\right)ue^{\frac{i}{\hbar}\boldsymbol{pr}}dV =$$

$$= -Zec\sqrt{\frac{2\pi\hbar}{\omega V}}\frac{r_0}{q}\int_V u'^*\left(\hat{\boldsymbol{p}}\left(\frac{1}{r}\right)\boldsymbol{e}\right)e^{\frac{i}{\hbar}(\boldsymbol{p}-\boldsymbol{p}'-\boldsymbol{q})\boldsymbol{r}}udV = -Zec\sqrt{\frac{2\pi\hbar}{\omega V}}\frac{r_0}{q}\int_V u'^*\left(\hat{\boldsymbol{p}}\left(\frac{1}{r}\right)\boldsymbol{e}\right)udV$$

$$= -Zec\sqrt{\frac{2\pi\hbar}{\omega V}}\frac{r_0}{q}\int_V u'^*\left(\hat{\boldsymbol{p}}\left(\frac{1}{r}\right)\boldsymbol{e}\right)udV = -Zec\frac{r_0}{q}\sqrt{\frac{2\pi\hbar}{\omega V}}\left(\boldsymbol{p}_{p'p}\boldsymbol{e}\right)$$

(6.5.2)

where we take into account the condition $\boldsymbol{e} \perp \boldsymbol{r}$, since in (6.5.2) transition to a radius-vector of the electron is carried out really only before collecting of exponents. It is - the law of conservation of the electron impulse; is a matrix element of the electron impulse in a field of a nucleus, so $\boldsymbol{p}_{p'p} = \int_V u'^{*}\hat{\boldsymbol{p}}\left(\frac{1}{r}\right)udV$. Wave functions of the electron before and after scattering are accepted as and $\varphi_{p'} = u'e^{\frac{i}{\hbar}\boldsymbol{p}'\boldsymbol{r}}$, where u and are amplitudes of wave functions. In transformations (6.5.2) the formula of the vector analysis [21] is used.

For a matrix element of the electron impulse we have:

$$\begin{aligned}
\boldsymbol{p}_{p'p} &= \hat{\boldsymbol{p}}_{p'p}\left(\frac{1}{r}\right) = -i\hbar\left(\nabla\frac{1}{r}\right)_{p'p} = -i\hbar\int_V \varphi^{*}_{p'}\left(\nabla\frac{1}{r}\right)\varphi_p dV = \\
&= -i\hbar\int_V u'^{*}e^{-\frac{i}{\hbar}\boldsymbol{p}'\boldsymbol{r}}\left(\nabla\frac{1}{r}\right)ue^{\frac{i}{\hbar}\boldsymbol{p}\boldsymbol{r}}dV = -i\hbar\int_V u'^{*}\left(\nabla\frac{1}{r}\right)ue^{\frac{i}{\hbar}(\boldsymbol{p}-\boldsymbol{p}')\boldsymbol{r}}dV = \\
&= -i\hbar\int_V u'^{*}\left(\nabla\frac{1}{r}\right)ue^{\frac{i}{\hbar}\boldsymbol{q}\boldsymbol{r}}dV = -i\hbar u'^{*}u\int_V\left(\nabla\frac{1}{r}\right)e^{\frac{i}{\hbar}\boldsymbol{q}\boldsymbol{r}}dV
\end{aligned} \tag{6.5.3}$$

Let's find integral [4, 53]:

$$\begin{aligned}
&\int_V\left(\nabla\frac{1}{r}\right)e^{\frac{i}{\hbar}\boldsymbol{q}\boldsymbol{r}}dV = \int_0^\infty\left(\nabla\frac{1}{r}\right)r^2dr\int_0^\pi e^{\frac{i}{\hbar}qr\cos\Theta}\sin\Theta d\Theta\int_0^{2\pi}d\psi = \\
&= -2\pi\int_0^\infty\left(\nabla\frac{1}{r}\right)r^2dr\left(\frac{\hbar}{iqr}\right)\int_0^\pi e^{\frac{i}{\hbar}qr\cos\Theta}d\left(\frac{i}{\hbar}qr\cos\Theta\right) = \\
&= -2\pi\int_0^\infty\left(\nabla\frac{1}{r}\right)r^2dr\left(\frac{\hbar}{iqr}\right)e^{\frac{i}{\hbar}qr\cos\Theta}\Bigg|_0^\pi = -2\pi\int_0^\infty\left(\nabla\frac{1}{r}\right)r^2dr\left(\frac{\hbar}{iqr}\right)\left(e^{-\frac{i}{\hbar}qr} - e^{\frac{i}{\hbar}qr}\right) = \\
&= 4\pi\hbar\int_0^\infty\left(\nabla\frac{1}{r}\right)\frac{\sin\frac{qr}{\hbar}}{qr}r^2dr = -4\pi\hbar\int_0^\infty\frac{1}{r^2}\frac{\boldsymbol{r}}{r}\frac{\sin\frac{qr}{\hbar}}{qr}r^2dr = -4\pi\hbar\int_0^\infty\frac{\boldsymbol{r}}{r}\frac{\sin\frac{qr}{\hbar}}{qr}dr = \\
&= -\frac{4\pi\boldsymbol{n}\hbar}{q}\int_0^\infty\frac{\sin\frac{qr}{\hbar}}{\frac{qr}{\hbar}}d\frac{qr}{\hbar} = -\frac{4\pi\boldsymbol{n}\hbar}{q}\int_0^\infty\frac{\sin x}{x}dx = -\frac{4\pi\boldsymbol{n}\hbar}{q}\frac{\pi}{2}
\end{aligned}$$

,

where is an integrated sine, - a unit vector in a direction of a radius-vector of the electron, figure 6.5.1. According to the law of energy conservation the module $q = k = \frac{\hbar\omega}{c}$.

Hence, the matrix element of an impulse of the electron in a nucleus field equals:

$$\boldsymbol{p}_{p^/ p} = -i\hbar u^{/*} u \int_V \left(\nabla \frac{1}{r} \right) e^{\frac{i}{\hbar}\boldsymbol{qr}} dV = iu^{/*} u \frac{4\pi\boldsymbol{n}\hbar^2}{q} \frac{\pi}{2}. \tag{6.5.4}$$

where it is accepted that the quantum starts up radial from atom. Thus the quantum has an impulse $\boldsymbol{k}$ which makes angle Θ with a vector $\boldsymbol{n}$, figure 6.5.1.

Assuming amplitudes of wave functions [4], we shall find:

$$\boldsymbol{p}_{p^/ p} = \frac{4\pi\boldsymbol{n}i\hbar^2}{Vq} = -\frac{4\pi\boldsymbol{q}i\hbar c}{Vq\omega}. \tag{6.5.5}$$

The probability of the electron's transition in unit of time with radiation of a photon in a solid angle $d\Omega$, taking into account (6.5.2), equals [4, 53]:

$$dW = \frac{2\pi}{\hbar} \left| H_{p^/ p} \right|^2 \rho(\omega) d\Omega = \frac{2\pi}{\hbar} \frac{2\pi\hbar}{\omega V} \frac{Z^2 e^2 c^2}{q^2} r_0^2 \left| \boldsymbol{p}_{p^/ p} \boldsymbol{e} \right|^2 \rho(\omega) d\Omega ,$$
$$= \frac{4\pi^2 Z^2 e^2 c^2}{\omega V q^2} r_0^2 \left| \boldsymbol{p}_{p^/ p} \boldsymbol{e} \right|^2 \rho(\omega) d\Omega = \frac{Z^2 e^2 \omega}{2\pi c q^2 \hbar} r_0^2 \left| \boldsymbol{p}_{p^/ p} \boldsymbol{e} \right|^2 d\Omega \tag{6.5.6}$$

where the number of photons in volume V, falling in unit power and angular intervals at the given photon polarization looks like [4].

To find the differential effective section of process with scattering of the electron in a spatial range of an impulse we use the formula [4, 18]:

$$d\sigma = \frac{dW}{j_0} \frac{V d^3 p^/}{(2\pi\hbar)^3}, \tag{6.5.7}$$

where is a density of a flux of falling electrons, N - number of falling electrons, v - their velocity [4, 21]. At this density of a flux equals $j_0 = \frac{v}{V}$.

Substituting (6.5.6) in (6.5.7), we shall find the differential effective section of the process of the electron transition $\boldsymbol{p} \to \boldsymbol{p}^/$:

$$d\sigma = \frac{Z^2 e^2 \omega V^2 m}{2\pi c(2\pi\hbar)^3 \hbar p q^2} r_0^2 \left| \boldsymbol{p}_{p'p} \boldsymbol{e} \right|^2 d\Omega d^3 p', \tag{6.5.8}$$

where it is taken into account that an impulse of the electron $p = mv$.

At the scattering of the electron in a solid angle do:

$$d^3 p' = p'^2 dp' do = p' d \frac{p'^2}{2} do = p' d \frac{p'^2 - p^2}{2} do = -p' \hbar m d\omega do, \tag{6.5.9}$$

where the law of energy conservation as and a condition are used. Substituting (6.5.9) in (6.5.8), and taking into account positive size $d\sigma$, we have:

$$d\sigma = \frac{Z^2 e^2 \omega V^2 m^2}{2\pi c(2\pi\hbar)^3 q^2} r_0^2 \left| \boldsymbol{p}_{p'p} \boldsymbol{e} \right|^2 \frac{p'}{p} d\Omega d\omega do. \tag{6.5.10}$$

Substituting (6.5.5) in (6.5.10), we shall transform the effective section of the process $\boldsymbol{p} \to \boldsymbol{p}'$:

$$d\sigma = \frac{Z^2 e^2 \omega V^2 m^2}{2\pi c(2\pi\hbar)^3 q^2} r_0^2 \left(\frac{4\pi\hbar c}{Vq\omega} \right)^2 \left| i\boldsymbol{qe} \right|^2 \frac{p'}{p} d\Omega d\omega do = \\ = \frac{Z^2 \alpha m^2 c^2}{\pi^2 q^4} r_0^2 \left| \boldsymbol{qe} \right|^2 \frac{p'}{p} d\Omega \frac{d\omega}{\omega} do \tag{6.5.11}$$

where is a fine-structure constant.

The received formula coincides with [18] where it is deduced by another, more formal method.

In a direction quantums do not take off, since in this case and $\boldsymbol{qe} = 0$. In a direction perpendicular $\boldsymbol{n}$ the probability of a start of quantum is maximal. Distribution of probability of a start of quantums it is shown on figure 6.5.1. The same conclusion is made in [46].

Let's lead integration on directions of a photon $\boldsymbol{k}$ and on directions of polarization. Assuming dipole character of quantums radiation, we have, that distribution of intensity of the taken off quantums submits to the law (1.3.5) or [21]. Integration on directions of polarization is reduced only to multiplication 2 since exists two directions of polarization. Taking into account $d\Omega = 2\pi \sin\Theta d\Theta$, we shall find:

$$d\sigma = 2\frac{Z^2\alpha m^2c^2}{\pi^2 q^4} r_0^2 |\boldsymbol{q}|^2 \frac{p'}{p}\left(\int_\Omega \sin^2\Theta d\Omega\right)\frac{d\omega}{\omega}do =$$
$$= 2\cdot 2\pi \frac{Z^2\alpha m^2c^2}{\pi^2 q^2} r_0^2 \frac{p'}{p}\left(\int_0^\pi \sin^3\Theta d\Theta\right)\frac{d\omega}{\omega}do = \qquad (6.5.12)$$
$$= 4\frac{Z^2\alpha m^2c^2}{\pi q^2} r_0^2 \frac{p'}{p}\left(\frac{\cos^3\theta}{3} - \cos\theta\right)_0^\pi \frac{d\omega}{\omega}do = \frac{16}{3}\frac{Z^2\alpha m^2c^2}{\pi q^2} r_0^2 \frac{p'}{p}\frac{d\omega}{\omega}do$$

Further we shall integrate on directions of electron scattering θ. Taking into account and $q^2 = p^2 + p'^2 - 2pp'\cos\theta$, we have:

$$d\sigma = \frac{32}{3} Z^2\alpha r_0^2 m^2c^2 \frac{p'}{p}\left(\int_0^\pi \frac{\sin\theta d\theta}{p^2 + p'^2 - 2pp'\cos\theta}\right)\frac{d\omega}{\omega} =$$
$$= \frac{32}{3}\frac{Z^2\alpha}{p^2} r_0^2 m^2c^2 \ln\frac{p+p'}{p-p'}\frac{d\omega}{\omega} = \frac{32}{3} Z^2\alpha r_0^2 \frac{m^2c^2}{p^2}\ln\frac{p+p'}{p-p'}\frac{d\omega}{\omega} \qquad (6.5.13)$$

The formula (6.5.13) characterizes spectral distribution of differential effective section of braking and scattering of electrons.

The differential effective section of electrons scattering is connected to numerical intensity of a falling flux of electrons under the formula $dP = I_0 d\sigma$, where is the differential of a flux of a scattering electrons (3.1.4). Taking into account, that each scattering electron generates one quantum of brake radiation, it is possible to assume, that is a numerical flux of quantums of brake radiation. Hence:

$$d\Phi_q = I_0 d\sigma = \frac{32}{3} I_0 Z^2\alpha r_0^2 \frac{m^2c^2}{p^2}\ln\frac{p+p'}{p-p'}\frac{d\omega}{\omega}. \qquad (6.5.14)$$

The energy flux of quantums of radiation equals:

$$d\Phi = \hbar\omega d\Phi_q = \frac{32}{3} I_0 Z^2\alpha\hbar r_0^2 \frac{m^2c^2}{p^2}\ln\frac{p+p'}{p-p'}d\omega. \qquad (6.5.15)$$

Hence, the spectral density of a photons flux does not depend on frequency that coincides with a conclusion [23].

Let's we note, that the received formula (6.5.15) does not take into account shielding of a nucleus field by atom's electrons. The account of effect of shielding is usually carried out by introduction empirical and half-empirical amendments [57].

Integrating (6.5.15) in limits from up to - the greatest possible frequency of a photon, we shall find dependence of a quantums flux of brake radiation on frequency:

$$\Phi = \frac{32}{3} I_0 Z^2 \alpha \hbar r_0^2 \frac{m^2 c^2}{p^2} \ln \frac{p+p'}{p-p'} (\omega_{max} - \omega) = \eta(\omega_{max} - \omega) \cdot \quad (6.5.16)$$

where is the factor dependent on a kind of a substance (charging number of nucleus Z), numerical intensity of a flux of electrons falling on substance, impulses of electrons before p and after interaction with nucleus. The maximal possible frequency of brake radiation can be found under the formula $\omega_{max} = \frac{E_0}{\hbar}$, where is the maximal energy of electrons in a spectrum β-particles. It is supposed, that all energy of the electron transform in energy of quantum of electromagnetic radiation.

If to accept kinetic energy of the falling electron as the value will represent a fraction of kinetic energy of the electron in its full energy.

Hence $\eta = \frac{16}{3} \frac{I_0}{\varepsilon} Z^2 \alpha \hbar r_0^2 \ln \frac{1+\chi}{1-\chi}$, where is the parameter describing efficiency of braking of electrons by a substance on one atom, - kinetic energy of the electron after interaction with atom. At braking full, at braking is absent. Size of linear energy transfer for electrons is $\text{LET} \sim E_{kin} - E'_{kin} = E_{kin}(1-\chi^2)$.

On figure 6.5.2 dependence of relative factor on parameter of efficiency of braking for various elements is shown. At small sizes of parameter of efficiency of braking dependence has practically linear character.

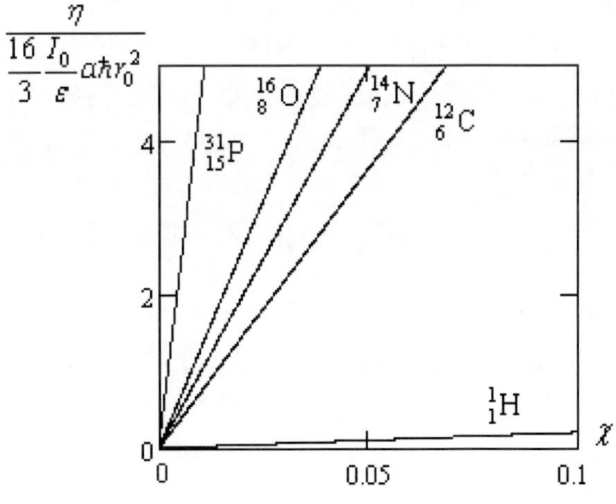

Figure 6.5.2.

6.5.2. The Relativistic Amendment

Let's consider influence of relativistic effects on character of brake radiation.

Let is the initial energy of the electron falling on the substance. There is a probability of a photon radiation at movement of the electron in a field of atom. The law of energy conservation looks like:

$$E^{/} + ck = E\,, \qquad (6.5.17)$$

where there is final energy of the electron.

During movement of the electron the transition of system atom-electron from an initial condition (start) in a final condition (finish) is carried out. Hamiltonian interactions equal:

$$H = H_{\text{int}} + V\,, \qquad (6.5.18)$$

where the Hamiltonian is conditionally separated into two addends. The first addend characterizes the radiation of the quantum by the electron, the second addend V - interaction of the electron with a field of atom.

In the case of Coulomb's field the energy of interaction of the electron and atom looks like:

$$V = \frac{Ze^2}{r}\,. \qquad (6.5.19)$$

Transition occurs through the intermediate states. Two types of transitions and, accordingly, two intermediate states I and II are possible.

1. At transition of system in the first intermediate state the electron with an impulse radiates a photon an impulse $\boldsymbol{k}$. After that the electron has an impulse $\boldsymbol{p}_{\text{I}}$. The law of an impulse conservation looks like:

$$\boldsymbol{p}_{\text{I}} = \boldsymbol{p} - \boldsymbol{k}\,. \qquad (6.5.20)$$

Transition is determined by addend in Hamiltonian (6.5.18). The square of the electron energy in the first intermediate state $E_{\text{I}}^{2} = p_{\text{I}}^{2}c^{2} + m_{0}^{2}c^{4}$, where is a mass of the rest of the electron.

Then the electron transitions into a final state $\text{I} \rightarrow F$. This transition is determined by the addend V in the Hamiltonian (6.5.18). The change of impulse in the electron looks like:

$$\boldsymbol{p}^{/} = \boldsymbol{p}_{\text{I}} - \boldsymbol{q}\,, \qquad (6.5.21)$$

where is a part of the electron impulse, transmitted to the atom in which there is an interaction.

2. If the system transitions into the second intermediate state $S \rightarrow \text{II}$, the electron, due to interaction with the atom, initially loses a part of its impulse. Its impulse becomes equal to:

$$\boldsymbol{p}_{\mathrm{II}} = \boldsymbol{p} - \boldsymbol{q}. \tag{6.5.22}$$

This transition is determined by the addend *V* in Hamiltonian (6.5.18).

The square of the electron energy in the second intermediate state $E_{\mathrm{II}}^2 = p_{\mathrm{II}}^2 c^2 + m_0^2 c^4$. Then, at the transition $\mathrm{II} \rightarrow F$, the electron radiates a photon an impulse ***k***, transitioning into a final state with an impulse $\boldsymbol{p}^{/}$:

$$\boldsymbol{p}^{/} = \boldsymbol{p}_{\mathrm{II}} - \boldsymbol{k}. \tag{6.5.23}$$

Transition is determined by addend in the Hamiltonian (6.5.18).

Combining (6.5.20) and (6.5.21) or (6.5.22) and (6.5.23), we notice that in both cases the conservation law of an impulse at transition looks like:

$$\boldsymbol{p} = \boldsymbol{p}^{/} + \boldsymbol{k} + \boldsymbol{q}, \tag{6.5.24}$$

i.e. the part of an initial impulse of the electron is transferred to a photon $\boldsymbol{k}$, and another part is transferred to the atom. Therefore, the size is identical to the first and second variants of transition.

Let's find a matrix element of that part of interaction of the electron with the atom which emits an impulse ***k*** to a photon. Research into the process of transition of the electron in the intermediate state $S \rightarrow \mathrm{I}$ is rather simple in connection with the linearity of the operator of Hamilton concerning the value $\left(\hat{p} - \frac{e}{c}\hat{A}\right)$, where ***p*** is an impulse of the electron, *e* - a charge of the electron, *c* - speed of light in vacuum, ***A*** is the vector-potential of a photon. At a transition we do not take into account the presence of the atom. With the use of the Coulombian calibrations of potentials $\mathrm{div}\boldsymbol{A} = 0$, we have [4, 21]:

$$\hat{H} = c\alpha\left(\hat{p} - \frac{e}{c}\hat{A}\right) + e\phi + m_0 c^2 \beta = c\alpha\hat{p} - e\alpha\hat{A} + e\phi + m_0 c^2 \beta, \tag{6.5.25}$$

where ϕ is the scalar potential of the photon, - the mass of the rest of the electron, - a matrix vector (a vector, whose coordinate projections are the square matrixes of the 4th order, with components $0, \pm 1, \pm i$), - a square matrix of the 4th order with components [4]. All multipliers of a matrix vector are the four-vectors, of which the fourth component is imaginary. The energy of interaction of the photon-electron characterizes only the second addend (6.5.25), i.e. the operator $\hat{H}_{\mathrm{int}} = -e\left(\alpha\hat{A}\right)$. In connection with that, the presence of the atom is not taken into account yet; in (6.5.25), as opposed to (6.5.1), it is supposed that the vector-potential of a photon is periodic, but not radiating the function of coordinates

$A=\sqrt{\frac{2\pi\hbar c^2}{\omega V}}\boldsymbol{e}e^{-\frac{i}{\hbar}\boldsymbol{kr}}$. Therefore, it is possible to write down the Hamiltonian of interactions with the radiated photon-electron as:

$$\hat{H}_{\text{int}}=-e\left(\boldsymbol{\alpha}\hat{\boldsymbol{A}}\right)=-e\sqrt{\frac{4\pi\hbar c^2}{\omega V}}(\boldsymbol{\alpha e})e^{-\frac{i}{\hbar}\boldsymbol{kr}}. \tag{6.5.26}$$

The matrix element of transition of the electron states looks like:

$$\begin{aligned}(H_{\text{int}})_{SI}&=\int\varphi_0\hat{H}_{\text{int}}\varphi_{\text{I}}^*dV=-e\sqrt{\frac{2\pi\hbar c^2}{\omega V}}\int\varphi_0(\boldsymbol{\alpha e})e^{-\frac{i}{\hbar}\boldsymbol{kr}}\varphi_{\text{I}}^*dV=\\&=-e\sqrt{\frac{2\pi\hbar c^2}{\omega V}}\int u_0e^{i\frac{\boldsymbol{pr}}{\hbar}}(\boldsymbol{\alpha e})e^{-\frac{i}{\hbar}\boldsymbol{kr}}u_{\text{I}}e^{-i\frac{\boldsymbol{p}_{\text{I}}\boldsymbol{r}}{\hbar}}dV=\\&=-e\sqrt{\frac{2\pi\hbar c^2}{\omega V}}\left[u_0(\boldsymbol{\alpha e})u_{\text{I}}^*\right]\int e^{-\frac{i}{\hbar}(\boldsymbol{p}_{\text{I}}-\boldsymbol{p}+\boldsymbol{k})\boldsymbol{r}}dV\end{aligned} \tag{6.5.27}$$

where is the amplitude of the wave function of the electron in an initial state of the electron. After radiation of a photon, we assume that the position of the electron has not changed; - amplitude of wave function in the intermediate state I, - the conjugate amplitude of this wave function.

Taking into account the conservation law of an impulse (6.5.20), we have $\int e^{-\frac{i}{\hbar}(\boldsymbol{p}_{\text{I}}-\boldsymbol{p}+\boldsymbol{k})\boldsymbol{r}}dV=V$. Assuming the normalizing volume $V=1$, we shall find a matrix element of transition $S\rightarrow\text{I}$:

$$(H_{\text{int}})_{SI}=-e\sqrt{\frac{2\pi\hbar^2c}{k}}\left[u_0(\boldsymbol{\alpha e})u_{\text{I}}^*\right]. \tag{6.5.28}$$

Let's find a matrix element of that part of interaction of the electron with the atom which occurs without the emission of a photon. We shall assume that the energy of interaction of the electron and atom has an essential value only near to atom; i.e. at r close to zero. At a matrix element is $(V)_{\text{IF}}\rightarrow 0$. This position, actually Born's approximation, will be taken into account at the further deduction. Far from the atom wave functions of the electron are flat waves and $\varphi=ue^{i\frac{\boldsymbol{p}'\boldsymbol{r}}{\hbar}}$, where u is the amplitude of the wave function of the electron in a final state.

$$
\begin{aligned}
(V)_{\mathrm{IF}} &= \int \varphi_{\mathrm{I}} \hat{V} \varphi^{*} dV = \int \varphi_{\mathrm{I}} \frac{Ze^2}{r} \varphi^{*} dV = \int u_{\mathrm{I}} e^{i\frac{\boldsymbol{p}_{\mathrm{I}}\boldsymbol{r}}{\hbar}} \frac{Ze^2}{r} u^{*} e^{-i\frac{\boldsymbol{p}'\boldsymbol{r}}{\hbar}} dV = \\
&= \left(u_{\mathrm{I}} u^{*}\right) \int \frac{Ze^2}{r} e^{i\frac{\boldsymbol{qr}}{\hbar}} dV = \left(u_{\mathrm{I}} u^{*}\right) \int_0^{\infty} \frac{Ze^2}{r} r^2 dr \int_0^{\pi} e^{i\frac{qr}{\hbar}\cos\theta} \sin\theta d\theta \int_0^{2\pi} d\psi = \\
&= 2\pi \left(u_{\mathrm{I}} u^{*}\right) \int_0^{\infty} \frac{Ze^2}{r} r^2 dr \int_0^{\pi} e^{i\frac{qr}{\hbar}\cos\theta} \sin\theta d\theta = 2\pi \left(u_{\mathrm{I}} u^{*}\right) \int_0^{\infty} \frac{Ze^2}{r} r^2 dr \frac{\hbar}{iqr} \left(e^{i\frac{qr}{\hbar}} - e^{-i\frac{qr}{\hbar}} \right) = \\
&= 2\pi \left(u_{\mathrm{I}} u^{*}\right) \int_0^{\infty} Ze^2 dr \frac{\hbar}{iq} \left(e^{i\frac{qr}{\hbar}} - e^{-i\frac{qr}{\hbar}} \right) = 2\pi \left(u_{\mathrm{I}} u^{*}\right) \int_0^{\infty} Ze^2 dr \frac{\hbar}{iq} 2i \sin\left(\frac{qr}{\hbar}\right) = \\
&= 4\pi \left(u_{\mathrm{I}} u^{*}\right) \frac{\hbar}{q} Ze^2 \int_0^{\infty} \sin\left(\frac{qr}{\hbar}\right) dr = -4\pi \left(u_{\mathrm{I}} u^{*}\right) \frac{\hbar^2}{q^2} Ze^2 \cos\left(\frac{qr}{\hbar}\right)\Bigg|_0^{\infty} = \frac{4\pi\hbar^2 Ze^2}{q^2} \left(u_{\mathrm{I}} u^{*}\right)
\end{aligned} \tag{6.5.29}
$$

where in this case angle θ is between the radius-vector of the electron $\boldsymbol{r}$ and a vector of an impulse $\boldsymbol{q}$, transferred to a nucleus. The law of conservation of an impulse (6.5.21) is used. At integration (6.5.29) the transition to spherical coordinates is lead. Oscillations of a matrix element far from atom at $r \to \infty$, as they do not have a physical sense, can be neglected.

At transition in an intermediate state II on non-radiating stage the matrix element of transition with the account (6.5.22) looks like:

$$
\begin{aligned}
(V)_{\mathrm{SII}} &= \int \varphi_0 \hat{V} \varphi_{\mathrm{II}}^{*} dV = \int u_0 e^{i\frac{\boldsymbol{pr}}{\hbar}} \frac{Ze^2}{r} u_{\mathrm{II}}^{*} e^{-i\frac{\boldsymbol{p}_{\mathrm{II}}\boldsymbol{r}}{\hbar}} dV = \left(u_0 u_{\mathrm{II}}^{*}\right) \int \frac{Ze^2}{r} e^{i\frac{\boldsymbol{qr}}{\hbar}} dV = \\
&= \left(u_0 u_{\mathrm{II}}^{*}\right) \int_0^{\infty} \frac{Ze^2}{r} r^2 dr \int_0^{\pi} e^{i\frac{qr}{\hbar}\cos\theta} \sin\theta d\theta \int_0^{2\pi} d\psi = \\
&= 2\pi \left(u_0 u_{\mathrm{II}}^{*}\right) \int_0^{\infty} \frac{Ze^2}{r} r^2 dr \int_0^{\pi} e^{i\frac{qr}{\hbar}\cos\theta} \sin\theta d\theta = \\
&= -2\pi \left(u_0 u_{\mathrm{II}}^{*}\right) \int_0^{\infty} \frac{Ze^2}{r} r^2 dr \frac{\hbar}{iqr} \left(e^{-i\frac{qr}{\hbar}} - e^{i\frac{qr}{\hbar}} \right) = \\
&= 2\pi \left(u_0 u_{\mathrm{II}}^{*}\right) \int_0^{\infty} Ze^2 dr \frac{\hbar}{iq} \left(e^{i\frac{qr}{\hbar}} - e^{-i\frac{qr}{\hbar}} \right) = 2\pi \left(u_0 u_{\mathrm{II}}^{*}\right) \int_0^{\infty} Ze^2 dr \frac{\hbar}{iq} 2i \sin\left(\frac{qr}{\hbar}\right) = \\
&= 4\pi \left(u_0 u_{\mathrm{II}}^{*}\right) \frac{\hbar}{q} Ze^2 \int_0^{\infty} \sin\left(\frac{qr}{\hbar}\right) dr = -4\pi \left(u_0 u_{\mathrm{II}}^{*}\right) \frac{\hbar^2}{q^2} Ze^2 \cos\left(\frac{qr}{\hbar}\right)\Bigg|_0^{\infty} = \frac{4\pi\hbar^2 Ze^2}{q^2} \left(u_0 u_{\mathrm{II}}^{*}\right)
\end{aligned} \tag{6.5.30}
$$

where is the conjugate amplitude of the wave function of the electron in an intermediate state II.

To radiating stage there corresponds a matrix element of transition:

$$\left(H_{\text{int}}\right)_{\text{II}F} = \int \varphi_{\text{II}} \hat{H}_{\text{int}} \varphi^{*} dV = -e\sqrt{\frac{2\pi\hbar c^2}{\omega V}} \int u_{\text{II}} e^{i\frac{\boldsymbol{p}_{\text{II}}\boldsymbol{r}}{\hbar}} (\boldsymbol{ae}) e^{-\frac{i}{\hbar}\boldsymbol{kr}} u^{*} e^{-i\frac{\boldsymbol{p}'\boldsymbol{r}}{\hbar}} dV =$$
$$= -e\sqrt{\frac{2\pi\hbar c^2}{\omega V}} \left[u_{\text{II}} (\boldsymbol{ae}) u^{*}\right] \int e^{\frac{i}{\hbar}(\boldsymbol{p}_{\text{II}} - \boldsymbol{p}' - \boldsymbol{k})\boldsymbol{r}} dV = -e\sqrt{\frac{2\pi\hbar^2 c}{k}} \left[u_{\text{II}} (\boldsymbol{ae}) u^{*}\right]$$

(6.5.31)

Thus, the law of conservation of an impulse (6.5.23) is used.

The compound matrix element of interaction of the electron and atom looks like:

$$\left(H_{\text{int}}\right)_{SF} = \frac{\left(H_{\text{int}}\right)_{S\text{I}} (V)_{\text{I}F}}{E - ck - E_{\text{I}}} + \frac{(V)_{S\text{II}} \left(H_{\text{int}}\right)_{\text{II}F}}{E - E_{\text{II}}}. \qquad (6.5.32)$$

In denominators of the addends (6.5.32) should be differences between the energy of an initial state of system of the electron-atom and energy; this system in intermediate states I and II.

However, in connection with that energy of the atom at its interaction with the electron is assumed to be constant, at the subtraction in the denominator's energy of atom disappears and it is not possible to take it into account. Therefore, only the denominator's energy of the electron and a photon is taken into account.

The intermediate state of the electron arises after the emission of a photon, but before the interaction with the atom; therefore, in a denominator of the first addend the energy emitted from the photon is subtracted. The intermediate state of the electron arises due to its interaction with the atom before the emission of a photon. Therefore, in a denominator of the second addend, it is not necessary to subtract the energy of a photon.

Taking into account (6.5.17), we have:

$$\left(H_{\text{int}}\right)_{SF} = \frac{\left(H_{\text{int}}\right)_{S\text{I}} (V)_{\text{I}F}}{E' - E_{\text{I}}} + \frac{(V)_{S\text{II}} \left(H_{\text{int}}\right)_{\text{II}F}}{E - E_{\text{II}}}. \qquad (6.5.33)$$

Let's substitute in (6.5.33) formulas (6.5.28) - (6.5.31):

$$\left(H_{\text{int}}\right)_{SF} = -e\sqrt{\frac{2\pi\hbar^2 c}{k}} \frac{4\pi\hbar^2 Z e^2}{q^2} \left(\frac{\left[u_0 (\boldsymbol{ae}) u_{\text{I}}^{*}\right]\left(u_{\text{I}} u^{*}\right)}{E' - E_{\text{I}}} + \frac{\left(u_0 u_{\text{II}}^{*}\right)\left[u_{\text{II}} (\boldsymbol{ae}) u^{*}\right]}{E - E_{\text{II}}} \right)$$

(6.5.34)

The probability of transition of the system in unit of time is proportional to a square of a corresponding matrix element [4]:

$$dW = \frac{2\pi}{\hbar}\left|\left(\hat{H}_{\text{int}}\right)_{SF}\right|^2 d\rho_F, \tag{6.5.35}$$

where is the number of final power states of the system, a unit interval of falling energy of a final state [46]. At the final state the electron with an impulse $\boldsymbol{p}^{/}$, with energy and a photon with an impulse $\boldsymbol{k}$ are present.

If is the number of power states of the radiated photon, the number of final states of the system equals:

$$d\rho_F = d\rho_E d\rho_k, \tag{6.5.36}$$

where is the number of power states of the electron [1]. It is taken into account that the energy of the electron $E^{/2} = p^{/2}c^2 + m_0^2c^4$; hence, and the portion of electrons emitted in a solid angle equals $\frac{do}{4\pi}$. The density of energies' states of electrons in the unit interval of energy in unit of volume equals $d\rho_E = \frac{p^{/}E^{/}do}{(2\pi\hbar)^3 c^2}$.

The number of power states of the radiated photon in unit of the volume $dV = 1$, radiated in a solid angle $d\Omega$; we find similarly [1].

The factor 2 means possibly two polarizations of a photon.

Hence, the number of final states of the system equals:

$$d\rho_F = 2\frac{p^{/}E^{/}k^2 dk\, do\, d\Omega}{(2\pi\hbar)^6 c^2}. \tag{6.5.37}$$

Let's substitute (6.5.37) in (6.5.35):

$$dW = \frac{2\pi}{\hbar}\left|\left(\hat{H}_{\text{int}}\right)_{SF}\right|^2 2\frac{p^{/}E^{/}k^2 dk\, do\, d\Omega}{(2\pi\hbar)^6 c^2}, \tag{6.5.38}$$

To find the differential effective section of process we shall divide the probability of transition in unit of time (6.5.38) on the numerical density of a flux of falling electrons [4, 18]. One electron is considered; therefore, we assume that in normalizing volume V there is only one electron $N = 1$. Therefore, the numerical density of the electron flux under the condition $V = 1$, equals $j_0 = nv = \frac{N}{V}v = v = \frac{mvc^2}{mc^2} = \frac{pc^2}{E}$, where m is the mass of the moving electron. Hence, the differential effective section of the process of interaction of the electron and atom looks like:

$$d\sigma_r = \frac{dW}{j_0} = \frac{2\pi}{\hbar}\left|\left(\hat{H}_{\text{int}}\right)_{SF}\right|^2 2\frac{p^/ E^/ Ek^2 dkdod\Omega}{p(2\pi\hbar)^6 c^4}, \tag{6.5.39}$$

Substituting in (6.5.39) formula (6.5.34), we shall find:

$$d\sigma_r = 2\frac{2\pi}{\hbar}\left(e\sqrt{\frac{2\pi\hbar^2 c}{k}}\frac{4\pi\hbar^2 Ze^2}{q^2}\right)^2 \frac{p^/ E^/ Ek^2 dkdod\Omega}{p(2\pi\hbar)^6 c^4}\delta = \\ = 2\frac{\alpha e^4 Z^2}{\pi^2 c^2}\frac{p^/ E^/ Ekdkdod\Omega}{pq^4}\delta, \tag{6.5.40}$$

where a fine-structure constant is $\alpha = \frac{e^2}{\hbar c} \approx \frac{1}{137}$. In (6.5.40) it is designated:

$$\delta = \left(\frac{\left[u_0(\alpha\boldsymbol{e})u_{\text{I}}^*\right]\left(u_{\text{I}}u^*\right)}{E^/ - E_{\text{I}}} + \frac{\left(u_0 u_{\text{II}}^*\right)\left[u_{\text{II}}(\alpha\boldsymbol{e})u^*\right]}{E - E_{\text{II}}}\right)^2 \tag{6.5.41}$$

Let's find the ratio of the differential effective section of the process in relativistic (6.5.40) and non-relativistic (6.5.11) variants:

$$\frac{d\sigma_r}{d\sigma} = 2\frac{\alpha e^4 Z^2}{\pi^2 c^2}\frac{p^/ E^/ Ekdkdod\Omega}{pq^4}\delta \cdot \frac{1}{\frac{Z^2 \alpha m^2 c^2}{\pi^2 q^4} r_0^2 |\boldsymbol{qe}|^2 \frac{p^/}{p} d\Omega \frac{d\omega}{\omega} do} = \\ = 2\frac{e^4}{m^2 c^4}\frac{E^/ Ek^2}{r_0^2}\cdot\frac{\delta}{|\boldsymbol{qe}|^2} = 2E^/ Ek^2 \cdot \frac{\delta}{|\boldsymbol{qe}|^2} = E^/ E \cdot \delta, \tag{6.5.42}$$

where is the classical radius of the electron. Integration on directions of polarization is reduced to multiplication on 2 since there exist two directions of polarization and $|\boldsymbol{q}| = |\boldsymbol{k}|$. Hence, taking into account a designation of value δ (6.5.41), we have:

$$d\sigma_r = d\sigma E' E\left(\frac{\left[u_0(\boldsymbol{\alpha e})u_{\mathrm{I}}^*\right]\left(u_{\mathrm{I}}u^*\right)}{E' - E_{\mathrm{I}}} + \frac{\left(u_0 u_{\mathrm{II}}^*\right)\left[u_{\mathrm{II}}(\boldsymbol{\alpha e})u^*\right]}{E - E_{\mathrm{II}}}\right)^2 =$$

$$= d\sigma E' E\left(\frac{\left[u_0(\boldsymbol{\alpha e})u^*\right]}{E' - E_{\mathrm{I}}} + \frac{\left[u_0(\boldsymbol{\alpha e})u^*\right]}{E - E_{\mathrm{II}}}\right)^2 = \tag{6.5.43}$$

$$= d\sigma E' E\left[u_0(\boldsymbol{\alpha e})u^*\right]^2\left(\frac{1}{E' - E_{\mathrm{I}}} + \frac{1}{E - E_{\mathrm{II}}}\right)^2$$

As the electrons' energy in an ultrarelativistic variant is very great, it is possible to use conditions and $E >> E_{\mathrm{II}}$. In this case:

$$E'E\left(\frac{1}{E' - E_{\mathrm{I}}} + \frac{1}{E - E_{\mathrm{II}}}\right)^2 \approx E'E\left(\frac{1}{E'} + \frac{1}{E}\right)^2 = \left(\sqrt{\frac{E}{E'}} + \sqrt{\frac{E'}{E}}\right)^2 \approx,$$

$$\approx \left(\sqrt{\frac{p}{p'}} + \sqrt{\frac{p'}{p}}\right)^2 = \frac{p}{p'} + \frac{p'}{p} + 2$$

where it is assumed in the ultrarelativistic variant and hence, $E \approx pc$, and $E' \approx p'c$.

Hence, the formula (6.5.43) will be transformed:

$$d\sigma_r = \left(\frac{p}{p'} + \frac{p'}{p} + 2\right)\left[u_0(\boldsymbol{\alpha e})u^*\right]^2 d\sigma, \tag{6.5.44}$$

Taking into account [46] and at $V = 1$, see paragraph 6.5.1 formula (6.5.5), we have:

$$d\sigma_r = \frac{4}{\pi^2}\left(\frac{p}{p'} + \frac{p'}{p} + 2\right)d\sigma = \gamma d\sigma, \tag{6.5.45}$$

Hence, for the account of the ultrarelativistic effects, it is necessary to multiply the differential effective section of the process in the non-relativistic variant on value $\gamma = \frac{4}{\pi^2}\left(\frac{p}{p'} + \frac{p'}{p} + 2\right)$.

In figure 6.5.3 the dependence of the relative factor on the parameter of efficiency of braking for various elements in the ultrarelativistic case is shown.

Let's consider in more detail used conditions and $E >> E_{II}$. We assume also that the quantum concerning low-energy is radiated. Hence, there is $E >> ck$. The first condition gives:

$$(E - ck)^2 >> p_I^2 c^2 - m_0^2 c^4; \quad E^2 >> p_I^2 c^2 - m_0^2 c^4;$$
$$p^2 c^2 + m_0^2 c^4 >> p_I^2 c^2 + m_0^2 c^4; \quad \left(p^2 - p_I^2\right)c^2 >> 0$$

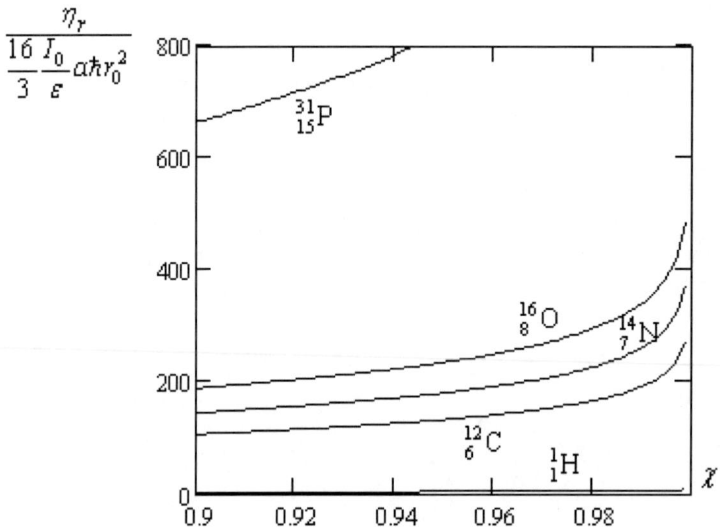

Figure 6.5.3.

Taking into account $\boldsymbol{p}_I = \boldsymbol{p} - \boldsymbol{k}$, we have $p_I^{\,2} = p^2 + k^2 - 2pk\cos\theta_0$, where is an angle between the initial direction of the electron and the radiated photon.

Hence:

$$\left(2pk\cos\theta_0 - k^2\right)c^2 >> 0; \quad 2pk\cos\theta_0 >> k^2; \quad 2p\cos\theta_0 >> k; \qquad (6.5.46)$$

The condition $E >> E_{II}$, taking into account (6.5.17), gives:

$$E^{\,2} >> E_{II}^{\,2}; \quad \left(E' + ck\right)^2 >> E_{II}^{\,2}; \quad E'^2 + 2E'ck + c^2k^2 >> E_{II}^{\,2};$$
$$p'^2 c^2 + m_0^2 c^4 + 2E'ck + c^2k^2 >> p_{II}^2 c^2 + m_0^2 c^4;$$
$$\left(p'^2 - p_{II}^2\right)c^2 + 2E'ck + c^2k^2 >> 0$$

Taking into account $\boldsymbol{p}_{\mathrm{II}} = \boldsymbol{p}^{/} + \boldsymbol{k}$, we have $p_{\mathrm{II}}{}^{2} = p^{/2} + k^{2} - 2p^{/}k\cos\theta$, where is an angle between the final direction of the electron and an impulse of a photon.

Hence,

$$\left(2p^{/}k\cos\theta - k^{2}\right)c^{2} + 2E^{/}ck + c^{2}k^{2} >> 0 \quad ;2p^{/}kc^{2}\cos\theta + 2E^{/}ck >> 0; \qquad (6.5.47)$$
$$p^{/}c\cos\theta + E^{/} >> 0;$$

Thus, conditions and at are equivalent to conditions or and $p^{/}c\cos\theta + E^{/} >> 0$.

The condition does not need to be discussed. The first condition specifies that at $p >> k$, or that also $E >> ck$, the cosine of an angle between the initial direction of the electron and the radiated photon is $\cos\theta_0 >> 0$, i.e. the approach to the unit and an angle $\theta_0 \to 0$. Thus, in the ultrarelativistic variant the radiation of photons is directed approximately along the initial direction of movement of the electron falling on the substance that corresponds with a conclusion [46].

Thus, at the braking of electrons of the β-rays in the substance there is a brake X-ray radiation.

At the analysis of the influence of β - rays on the substance, it is necessary to take into account the fact that arising brake ionizing electromagnetic radiation acts on a rather large distance from a trajectory of movement of the electron.

For the electrons concerning small energies the brake radiation is directed perpendicularly on planes of the electrons' scattering, and for the electrons with large energy the brake radiation is directed approximately along a direction of electrons falling on the substance.

6.5.3. Spectrum of the Brake Electromagnetic Radiation

For an estimation of the received results' correction it is necessary to compare it to known experimental data.

There is brake radiation in the X-ray range which is usually received in X-ray tubes. The electrons are accelerated in the space between the cathode and the anode. Between the cathode and the anode there is also accelerating voltage. At the braking in substance of the anode, the electrons cause brake electromagnetic radiation from the anode.

A comparison of the formula (6.5.16) with the results of the experiments [23] shows that it does not correspond with the experiments on two parameters. First, the dependence of an energy flux of quantums of brake radiation on charging number Z of atoms of braking substance is not square-law, but linear. Secondly, the frequency dependence, on the contrary, is not linear but square-law. According to [23], it is basically caused by two factors. A portion of the electrons is captured by atoms into stationary orbits, and also there is the probability of the return diffusion of electrons from the braking substance. These phenomena were not taken into account in our analysis. At the capture of electrons into a stationary orbit, it stops to participate in the creation of quantums of brake radiation. Therefore, the flux of energy of the quantums is reduced. The effective section of electron capture by the atom proportionally to its charging number $\sigma_a \sim Z$. The flux of energy of brake electromagnetic radiation is inversely proportional to the effective section of electron capture $\Phi \sim \dfrac{1}{\sigma_a}$; therefore, linear dependence is observed.

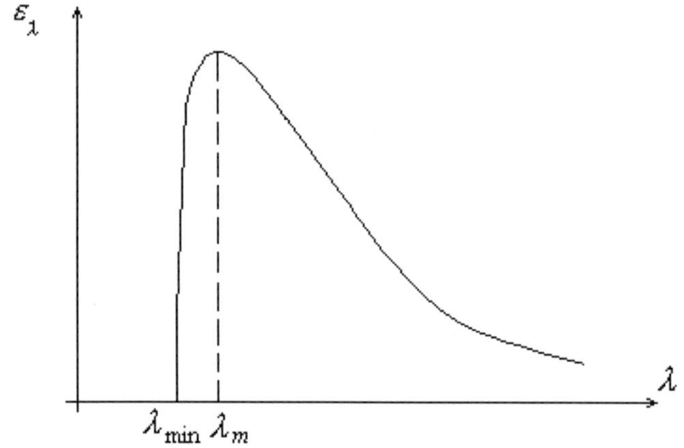

Figure 6.5.4.

The intensity of return diffusion of electrons from the anode depends on their energy. Electrons with lower energy will penetrate into the substance of the anode less deeply than electrons with higher energy. Therefore, the electrons with lower energy will take off back from the anode more often and participate less in the creation of brake electromagnetic radiation. This results in a reduction in a spectrum of brake radiation of a fraction of low-frequency quantums. There is a nonlinear dependence of an energy flux of brake radiation on the frequency of quantums.

According to results of experiments, the energy flux of brake electromagnetic radiation in the X-ray tube equals:

$$\Phi = k_1 Z I (\omega_{\max} - \omega)^2, \tag{6.5.48}$$

where I is a current in the X-ray tube, determined by the electrons' flux from the cathode to the anode; is constant.

Carrying out simple transformations, with the account $\omega = \dfrac{2\pi c}{\lambda}$, we find the dependence of spectral density of the energy of brake electromagnetic radiation on wave length of this radiation:

$$\varepsilon_\lambda = \frac{d\Phi}{d\lambda} = k_2 Z I \left(\frac{\lambda - \lambda_{\min}}{\lambda^3 \lambda_{\min}} \right), \tag{6.5.49}$$

where is a constant, - the shortest wave length in a spectrum of brake radiation.

It is possible to find the connection of the wave length on which there is a maximum of energy of brake radiation and the shortest wave length from the extreme condition of the function of spectral density of energy $\dfrac{d\varepsilon_\lambda}{d\lambda} = 0$.

In figure 6.5.4 the approximate form of dependence of spectral density of the energy of brake electromagnetic radiation from wave length is shown, according to the formula (6.5.49).

CONCLUSION

Consideration of the problem of interaction of an electromagnetic field with a substance shows that there are many questions to which physical science has not given an answer. The physics of the first millenium was inspired absolutely by other ideas, paradigms and methodology. It is enough to read Aristotle's book 'Physics'. But his words "We then are sure in knowledge of anything when we find out its first reasons, the first beginnings and we decompose it down to elements" will apparently never become outdated.

The beginning of the 20th century was marked by two scientific revolutions: the creation of quantum mechanics and the theory of relativity; special and general. These scientific revolutions have entered new concepts and ideas into physical science, allowing us to make various scientific-technological achievements; to create lasers, to use atomic energy, to start to study the depths of a microcosm, etc. The emotional echo of that time is the Nobel Prize which is the most prestigious award for the scientist today. Its role in the beginning of XX century was a little different from now. The governments of different countries at that time awarded officials, diplomats and military men with honors . The Nobel Prize, actually, has included scientists in its list of recipients, lifting the prestige of science as a whole.

Unfortunately, unlike biology, achievements in physics have remained on one level since the beginning of the 20th century. The development of physics was basically in the evolutionary stages, although specific steps of progress took place.

The mathematical methods used in physics were always the function of an epoch. Mathematics develops basically due to internal incentives. Physics uses various mathematical methods, frequently proceeding not from requirements of physics, and following personal preferences, styles of thinking, received education, etc. A vivid example is the practically simultaneous research of the electromagnetic induction law by Faraday and Maxwell. With respect to the distinction of the used mathematical methods, it is impossible to count the research of one scientific work as being more valuable than the other.

In this book we adhered to a golden rule; not overloading it with excessive mathematical transformations, not making attempts at the deliberate application of the latest achievements of mathematics. In using mathematics, the purpose was evidence and a transparency of deductions; the absence of mathematical half-words. The focus, to which was paid special attention, was the disclosing of the physical essence of the processes, open and fair discussion not ultimately clear, in the opinion of the author, at the given stage of development of a science.

Naturally, many questions have remained behind the frameworks of the book; for example, the quark theory of a matter structure. Here, we adhered to the position that, while this or that physical theory has not received practical application, instead of theoretical or even experimental acknowledgement, its statement in a similar book is not absolutely comprehensible.

Undoubtedly, new revolutions in physics wait ahead of us. They will also be followed with new unpredictable opportunities, but meanwhile these opportunities remain the subject of discussion among fantasts.

REFERENCES

[1] Javorsky B.M., Detlaf A.A. "Physical Handbook." Moscow, *Science*, 1990. 624 p.

[2] Feynman R.P., Leighton R.B., Sands M. "The Feynmam lectures on physics." V. 7. Moscow, *World*, 1978. 288 p.

[3] Korn G.A., Korn T.M. "Mathematical Handbook." Moscow: *Science*, 1968. 720 p.

[4] Levich V.G., Vdovin Yu.A., Myamlin V.A. "Theoretic Physics Course." V. 2. Moscow, *Phismatgis*, 1962. 820 p.

[5] Landsberg G.S. "Optics." Moscow, *Science* 1976, 928 p.

[6] Kostukov N.S., Lukichev A.A., Muminov M.I., etc. "ε and tanδ at an irradiation." Moscow, *Science*, 2002, 328 p.

[7] Berezovskij V.A., Kolotilov N.N. "Biophysical characteristics of the tissues of organism." Kiev, *Naukova dumka*, 1990, 224 p.

[8] Landau L.D., Lifshits E.M. "Electrodynamics of Continuous Environments." V. 8, Moscow, *PhysMathLit,* 2003, 656 p.

[9] Hippel A.R. "Dielectrics and waves." Moscow, *World*, 1960, 336 p.

[10] Katsenelenbaum B.Z., Korshunova E.N., Sivov A.N., Shatrov A.D. "Chirality electrodynamic objects." *Successes of physical sciences*, 1997, V. 167. No 11. pp. 1201-1212.

[11] Jatsyshen V.V., Dubovoj E.S. "Calculation of polarizing characteristics for the past and reflected electromagnetic waves from the layered chirality objects." *Physics of wave processes and radio engineering systems*, 2008, V. 11, No 3, pp. 44-53.

[12] Dodd R., Eilbek G., Ghibbon G., Morris G. "Solitons and Nonlinear Wave Equations." Moscow, *World*, 1988. 696 p.

[13] Ablovits M., Sigur H. "Solitons, methods of reverse task." Moscow: *World*, 1987. 480 p.

[14] Whitham G.B. "Linear and Nonlinear Waves." Moscow: *World*, 1977. 624 p.

[15] Lamb G. L. "Elements of soliton theory." Mogilov, *Bibfizmat*, 1997, 294 p.

[16] Volobuev A.N., Neganov V.A., Bolochagin V.J. "Interaction of solitones, described by the differential equation with logarithmic non-linearity." *Physics of wave processes and radio engineering systems*, 2000, V. 3, No 1, pp. 12-18.

[17] "Solitons." Ed. Bullough R. K., Caudrey P.G. Novokuznetsk, *Physical and mathematical institute*, 1999. 408 p.

[18] Berestetsky V.B., Lifshits E.M., Pitaevsky L.P. "Quantum electrodynamics." Moscow, *Science*, 1989, 728 p.

[19] Davidov A.S. "Quantum mechanics." Moscow, *Phismatgis*, 1963.748 p.
[20] Beernbaum J. "Optical quantum generators." Moscow, *Soviet radio*, 1967, 168 p.
[21] Levich V.G. "Theoretic Physics Course." V. 1. Moscow: *Phismatgis*, 1962. 696 p.
[22] Mihajlov A.I., Mihajlov I.A. "Double nuclear photoeffect in relativistic area. Angular and power distributions of photoelectrons." *JETP*, 1998. V. 114. Realise 5, pp. 1537-1554.
[23] Blohin M.A. "Physics of X-rays." Moscow, *Gostehizdat*, 1957, 518 p.
[24] Steigerwald D.A., Egelhoff W.F., jr. *Phys. Rev. Lett.* 1988, V. 60, p. 2558.
[25] Wichmann E.H. "Quanyum physics." Moscow, *Science*, 1977. 416 p.
[26] Kidd R., Ardini J., Anton A. "Compton Effect as a Duble Doppler Shift." *American Journal of Physics*, Vol. 53, No. 7, 1985, p. 641.
[27] Ashkroft N., Mermin N. "Physics of a solid body." V. 1, Moscow, *World*, 1979, 310 p.
[28] Lifshits E.M., Pitaevsky L.P. "Physical kinetics." V. 10. Moscow, *Fizmatlit*, 2002, 536 p.
[29] Vladimirov V.S. "Mathematical problems of the one-high-speed theory of carry of particles." *Works of mathematical institute of AN USSR*, 1961, V. 61. pp. 3 - 11.
[30] Bulyshev A.E., Preobrazhenskiy N.G., Suvorov A. E. "Carry of radiation to spectral lines." *Successes of physical sciences*, 1988, V. 156, Realise 1. pp. 154 - 176.
[31] Troshchiev V.E., Nifanova A.V., Troshchiev J.V. "The characteristic approach to approximation of laws of conservation in the kinetic equations of carry of radiations." *DAS RF*, 2004, V. 394, No 4, pp. 454 - 458.
[32] Kompaneets A.S. "About an establishment of thermal balance between quantums and electrons." *JETP,* 1956, V. 31, Realise 5, pp. 876 - 885.
[33] Zeldovich J.B., Levich E.V. "Bose - condensation and shock waves in a spectrum of photons." *JETP*, 1968, V. 55, Realise 6, pp. 2423 - 2429.
[34] Madelung O. "The theory of a solid body." Moscow, *Science*, 1980. 416 p.
[35] Amirov P.X., Gusjatnikov V.N. "Dynamic influence of a constant electric field on kinetics of the photons cooperating with electrons of the semiconductor." *Physics and technics of semiconductors*, 2001, V. 35, Realise 5, pp. 528 - 533.
[36] Matveev A.N. "Atom physics." Moscow, *Higher school*, 1989, 440 p.
[37] Matveev A.N. "Molecular physics." Moscow, *Higher school*, 1981, 400 p.
[38] Zeldovich J.B., Novikov I.D. "The theory of gravitation and evolution of stars." Moscow, *Science*, 1971, 484 p.
[39] Landau L.D., Lifshits E.M. "Statistical Physics." V. 5, Moscow, *Science,* 1976, 584 p.
[40] Lifshits E.M., Pitaevsky L.P. "Statistical Physics." P. 2, V. 9. Moscow, *Fizmatlit*, 2002, 494 p.
[41] Feynman R.P., Leighton R.B., Sands M. "The Feynmam lectures on physics." V. 8,9. Moscow, *World*, 1978. 528 p.
[42] Javorsky B.M., Detlaf A.A., Milkovsky L.B. "Physics Course." V. 1, Moscow, *Higher school*, 1965. 376 p.
[43] Born M., Wolf E. "Principles of Optics." Moscow, *Science,* 1973, 720 p.
[44] Landau L.D., Lifshits E.M. "The Theory of Field." V. 2, Moscow, *Science,* 1967, 460 p.
[45] Ashkroft N., Mermin N. "Physics of a solid body." V. 2, Moscow, *World*, 1979, 424 p.

[46] Gajtler V. "Quantum theory of radiation." Moscow, *In. Lit.*, 1956, 492 p.

[47] Mott M., Sneddon I. "Wave Mechanics and its Applications." Moscow, *KomKniga*, 2007, 432 p.

[48] Dirac P.A.M. "Recollections of an Exciting Era." Moscow, *Science*, 1990, 208 p.

[49] "Mathematical encyclopedia dictionary." Under ed. A.M. Prokhorov, Moscow, *Big Soviet encyclopedia*, 1995, p. 570.

[50] Tomonaga Sin-Itiro. "Development of Quantum Electrodynamics." Nobel speech. *Successes of physical sciences*, 1967, V. 91, Realise 1. pp. 61 - 69.

[51] Sushchinsky M.M. "Compelled scattering of light." Moscow, *Science*, 1985, 176 p.

[52] Volkenchtein M.V. "Biophysics." S-Petersburg, *Lan*, 2008, 576 p.

[53] Landay L.D., Lifshits E.M. "Quantum mechanics." V. 3, Moscow, *Fizmatlit*, 2004, 800 p.

[54] Condon E. "Theory of optical rotating ability." *Successes of physical sciences*, 1938, V. 19, Realise 3. pp. 380 - 430.

[55] Landau L.D., Lifshits E.M. "Mechanics." V. 1, Moscow, *Science*, 1988, 216 p.

[56] Djerassi C. "Optical Rotatory Dispersion." Moscow, *In. Lit.*, 1962, 300 p.

[57] Savinsky A.K. "Interaction of electrons with tissue-equivalent environments." Moscow, *EnergyAtomIzdat*, 1984, 112 p.

[58] "The State Pharmacopoeia of the USSR." 11 ed. Release 1, Moscow, *Medicine*, 1987, 336 p.

[59] Kudryashov J.B. (2004) "Radiating biophysics." Moscow, *Fizmatlit*, 2004, 446 p.

[60] "Physical encyclopedia dictionary." Under ed. A.M. Prokhorov, Moscow, *Soviet encyclopedia*, 1983, p. 377.

[61] Compton A.H. // *Bull. Nat. Res.. Conc.* 1922. V.4. Part 2. № 20. pp. 18–20.

[62] Rosenfeld L., *Z. Physik.*, 1928. V. 52. p. 161.

[63] Moffitt W., Moscowitz A., *J. Chem. Phys.* 1959. V. 30. p. 648.

[64] Kuhn W., *Z. Phys.Chem.*, 1929. B. 4. p. 14.

[65] Kuhn W., Freudenberg. *Hand- und Jahrbuch der chemmischen Physik*. 1932. V. 8. part 3. p. 59.

[66] Drude P. Lehrbuch der Optic. 2 Aufl., *S. Hirzel Verlag*. Leipzig, 1906.

INDEX

E

F

G

H

I

K

L

M

N

O

P

Q

R

S